D0914083

Combinatorial Algorithms

For Computers and Calculators

Second Edition

This is a volume in
COMPUTER SCIENCE AND APPLIED MATHEMATICS

A Series of Monographs and Textbooks

Editor: WERNER RHEINBOLDT

A complete list of titles in this series appears at the end of this volume.

Combinatorial Algorithms

For Computers and Calculators

Second Edition

ALBERT NIJENHUIS and HERBERT S. WILF

Department of Mathematics
University of Pennsylvania
Philadelphia, Pennsylvania

ACADEMIC PRESS, INC.

(Harcourt Brace Jovanovich, Publishers)

Orlando   San Diego   San Francisco   New York
London   Toronto   Montreal   Sydney   Tokyo

COPYRIGHT © 1978, BY ACADEMIC PRESS, INC.
ALL RIGHTS RESERVED.
NO PART OF THIS PUBLICATION MAY BE REPRODUCED OR
TRANSMITTED IN ANY FORM OR BY ANY MEANS, ELECTRONIC
OR MECHANICAL, INCLUDING PHOTOCOPY, RECORDING, OR ANY
INFORMATION STORAGE AND RETRIEVAL SYSTEM, WITHOUT
PERMISSION IN WRITING FROM THE PUBLISHER.

ACADEMIC PRESS, INC.
Orlando, Florida 32887

United Kingdom Edition published by
ACADEMIC PRESS, INC. (LONDON) LTD.
24/28 Oval Road, London NW1 7DX

Library of Congress Cataloging in Publication Data

Nijenhuis, Albert.
    Combinatorial algorithms for computers and calculators.

    (Computer science and applied mathematics)
    First ed. published in 1975 under title: Combina-
torial algorithms.
    Bibliography: p.
    Includes index.
    1. Combinatorial analysis--Computer programs.
2. Algorithms. I. Wilf, Herbert S., 1931-
joint author. II. Title.
QA164.N54   1978   511'.6'0285425   78-213
ISBN 0-12-519260-6

PRINTED IN THE UNITED STATES OF AMERICA

84 85 86 87   9 8 7 6 5 4 3

To

P. G. J. Vredenduin, for his early inspiration and support,
Henry and *Bernice Tumen*, with affection and esteem.

Contents

Preface to Second Edition

Since the appearance in 1975 of this work, the field of combinatorial algorithms has continued its rapid evolution. We have substantially rewritten several of the chapters in order to take account of theoretical or algorithmic improvements, and to clarify the presentation.

The result has been that a number of speedups, storage economies, and program simplifications have been made, some significant new theoretical material such as that in the two new chapters (13 and 14) has been added, and some minor errors have been corrected.

As an inducement to any reader who will point out to us an error or misprint, we offer a copy of the complete errata sheet as it stands at that time.

In the spring of 1977 one of us (H.W.) had the pleasure of teaching a course at Swarthmore College based on this book. One of the students in that class, Mr. David Bayer, made a number of original and insightful observations which have resulted in improvements to our algorithms. We acknowledge our debt to him and wish him every success in his young career as a mathematician.

Preface to First Edition

In the course of our combinatorial work over the past several years, we have been fond of going to the computer from time to time in order to see some examples of the things we were studying. We have built up a fairly extensive library of programs, and we feel that others might be interested in learning about the methods and/or use of the programs. This book is the result.

It can be read as a collection of mathematical algorithms, and as such we hope the reader will find much that is new and interesting. Yet to do so would be to miss something that to us seems essential: the interchange between the computer programs per se, the computer, the algorithms, and ultimately the mathematics. To capture the complete spirit of this work, we urge the reader to study the programs themselves. The extra dimension that the computer and the mathematics bestow on each other is, we believe, worth the effort. Above all, we hope we have placed in the reader's hands a kit of building blocks with which the reader can construct more elaborate structures of his or her own.

The second-named author expresses his appreciation to the John Simon Guggenheim Memorial Foundation for its support during the writing of this book, and to Rockefeller University of New York for its hospitality during the same period. Much of the original research described herein was supported by the National Science Foundation.

We wish to thank Donald E. Knuth for reading the manuscript and for making a number of extremely valuable suggestions that resulted in improvements.

Introduction

AIMS

This book can be read at several levels. Those whose only need is to use one of the computer programs can turn immediately to those pages and satisfy their wants. Thus, on one level, this is a collection of subroutines, in FORTRAN, for the solution of combinatorial problems.

At the other extreme, pure mathematicians with no need of computer programs will find much that is new and hopefully interesting in these pages. For example, in the special section *Deus ex Machina* (pp. 78–87), the random selection algorithms of Chapters 10, 12, and 29 are shown to be manifestations of a general phenomenon which sheds light on a number of seemingly unrelated threads of research in combinatorial analysis.

Between these two extremes is a rapidly growing category of (frequently youthful) persons who have access to a fancy calculator (hand-held or table-top). They may not be interested in either the detailed mathematics or the FORTRAN programs – yet we hope they will find much to stimulate them and help them prepare their own programs.

Our hope, however, is that many readers will want to follow the entire road from general mathematics to particular mathematics to informal algorithm to formal algorithm to computer program and back again, which occurs in virtually every chapter of the book.

Our other hope is that readers will view these methods and programs as a beginning set of building blocks for their own kit of tools and will go on to add to these tools to meet their own needs, so that the contents of this book will be not a collection of pretty artifacts to be looked at but basic elements of the growing and working equipment of scientific investigation and learning.

HIGHLIGHTS

We preview some of the features which lie ahead. First, concerning the random choice algorithms previously mentioned, in Chapter 10 there is an algorithm for selecting, at random, a partition of an integer n, so that all are equally likely to occur. This seems to work for a special reason, but actually it works for a very general reason described in the section *Deus ex Machina* which follows Chapter 10. Another outcropping of the same idea is found in Chapter 29 where we can select, at random, an unlabeled rooted tree on n vertices so that all are equally likely, and, in Chapter 12, a closely related idea results in the selection of a random partition of an n-set.

In Chapter 13 there appears a unified approach to the problem of selecting objects from combinatorial families. This approach may help to pull together the methods which appear in Chapters 1–12 as well as to motivate the interesting developments of Chapter 14, in which a random selection algorithm is used to *prove* a rather difficult theorem.

In Chapter 23 there is a calculation of the permanent function which for an $n \times n$ matrix is about $n/2$ times faster than standard algorithms at no cost in storage. It is in essence a variation of a known method in which subsets of a set are processed in a special sequence. The sequence is provided by the program in Chapter 1.

Further algorithms which merit special mention are the revolving-door method of Chapter 3, which extends the spirit of Chapter 1 into the realm of fixed cardinality, the sequential generation of compositions, permutations, and partitions of Chapters 5, 7, and 9, the random selection of k-subsets (Chapter 4) and compositions (Chapter 6), and the logarithmic-derivative-based composition of power series (Chapter 21).

Chapter 22 on network flows uses a new implementation of a standard algorithm and works very smoothly on graphs whose edges have positive capacities in both directions (e.g., undirected graphs). Applications include graph connectivity and various matching problems. The Möbius sequence (Chapters 24–26) is also elementary but minimizes storage space and computing time by suitable relabeling of elements.

The backtrack method of Chapter 27 is well known, and we have added nothing new except for the specific implementation and applications. The renumbering method of Chapter 17 is curiously arresting. The problem it solves is nonexistent within the scope of mathematics, which does not concern itself with duplication of storage requirements. Yet, in computation, such questions as this thrust themselves to the fore time and time again.

CATEGORIES OF USAGE (PART I)

We distinguish two kinds of usage, and try to deal with them both: the exhaustive search and the random sampling. In algorithms of search type, we have before us a list of combinatorial objects and we want to search the entire list, or perhaps to search sequentially until we find an object which meets certain conditions. For example, we may wish to hunt through the list of all 3,628,800 permutations of 10 letters in order to find the distribution of their largest cycles.

Random sampling, on the other hand, is done when we want to get the order of magnitude of a quantity of interest, but the exact determination of the quantity by exhaustive search would be so time consuming as to be impracticable. Thus, if we wanted to search through the 87,178,291,200 permutations of 14 letters to examine their largest cycles, it might be advisable to consider random sampling techniques.

These two categories of use call for different kinds of algorithms. For a search, we want a subprogram which, each time we call upon it, will present us with one of the objects on our list. We can then process the object and call the subprogram again to get the next object, etc. In broad outline, such a subprogram must (a) realize when it is being called for the first time, (b) remember enough about its previous output so that it can construct the next member of the list, (c) realize, at the end, that there are no more objects left, and (d) inform the calling program that the end has been reached. The algorithms in this work which are of the above type, and the lists of objects which they search sequentially are

(1) NEXSUB   All subsets of a set of n elements (n given).
      LEXSUB   All subsets of a set of n elements (n given).
(3) NEXKSB   All k-subsets of a set of n elements (n, k given).
      NXKSRD   All k-subsets of a set of n elements (n, k given).

(5) NEXCOM   All compositions of n into k parts (n, k given).
(7) NEXPER   All permutations of a set of n elements (n given).
(9) NEXPAR   All partitions of an integer n (n given).
(11) NEXEQU   All partitions of a set of n elements (n given).
(14) NEXYTB   All Young tableaux of a given shape.

The prefix NEX suggests "next," because these routines deliver the next subset, the next k-subset, etc.

In each case the routines are written in such a way as to minimize, if not actually to eliminate, the bookkeeping responsibilities of the calling program of the user. The detailed plan of construction of the programs will be discussed overall in the next section and individually in each chapter.

The logic of a random sampling subroutine is much simpler. Given the input parameters, the subprogram is expected to select at random just one object from the list specified by the input parameter. Here, "at random" has the strict and consistently followed interpretation that *each object on the list has equal a priori probability of being selected.* If the list is short, of course, one might consider constructing the whole list and selecting a member from it at random. The need for random selection methods, however, expresses itself at exactly the point where the above naïve approach fails, namely, when the lists are too long to deal with *in toto.* What are needed, therefore, are methods of *constructing* objects of desired type (the construction depending on the choices of random numbers) in such a way that all objects of the desired type are equally likely to result.

In some cases, these algorithms are trivial (Chapters 2 and 8); in other cases, a simple application of known theorems yields the algorithm (Chapter 24); and in still other cases, new methods are needed and the algorithms appear here for the first time (Chapters 10, 12, and 25). The complete list of algorithms in this work which are of this "random" type and the lists of objects from which they select are

(2) RANSUB   All subsets of an n-set (n given).
(4) RANKSB   All k-subsets of an n-set (n, k given).
(6) RANCOM   All compositions of n into k parts (n, k given).
(8) RANPER   All permutations of a set of n elements (n given).
(10) RANPAR   All partitions of the integer n (n given).
(12) RANEQU   All partitions of an n-set (n given).
(14) RANYTB   All Young tableaux of a given shape.
(24) RANTRE   All labeled trees on n vertices (n given).
(25) RANRUT   All rooted unlabeled trees on n vertices (n given).

STRUCTURE OF THE CHAPTERS

Each chapter follows roughly the same format, as detailed on page 5.

(a) The mathematical basis of the problem is examined and the chosen algorithm is informally described.

(b) A formal algorithm is stated.

(c) Where appropriate, a complete computer flow chart is shown in which the numbering of boxes mirrors the numbering of instructions in the actual computer program.

(d) The flow chart, if present, is described.

(e) Just prior to the FORTRAN program itself there appears a "Subroutine Specifications" list. Readers who want only to use a certain program should turn first to this page in the chapter, for it contains complete descriptions of the variables of the subroutine as a user would need to know them. This Specifications list is described in detail in the next section below.

(f) The FORTRAN program. All programs are written in SUBROUTINE form. While we have attempted to speak least-common-denominator FORTRAN, it cannot be expected that every program will always work with every compiler without slight changes. We believe that such changes will be minimal, and usually nonexistent.

(g) A sample problem, described in detail, followed by output reproduced from an actual machine run of the program.

THE SPECIFICATIONS LIST

Each computer program is immediately preceded by a specifications list which shows the name of the program and then the exact form of its calling statement. Next there is a capsule statement of what the program does, and then there is a list which shows, for *each variable which is named in the calling statement*, the following information:

(a) The name of the variable.

(b) The type of the variable; e.g. INTEGER, REAL(N), INTEGER(K), DOUBLE PRECISION(M,N), etc. If a parenthesis is present, the variable is an array, and the quantities inside the parentheses indicate the maximum size of the array, expressed in terms of SUBROUTINE parameters.

(c) The column headed *I/O/W/B* describes the role which is played by the variable in the interaction between the subroutine and the "outside world." In this column, opposite each variable, will be found one of the five designations: *I* (input), *O* (output), *I/O* (input–output), *W* (working), *B* (bookkeeping). We have found it desirable to give quite precise meanings to these designations according to the truth or falsity of the following three propositions:

P1: The value(s) of this variable at the time the subroutine is called affects the operation of the subroutine.

P2: The value(s) of this variable is changed by the operation of the subroutine.

P3: The computation of this variable is one of the main purposes of the subroutine.

Then, our precise definitions of the five designations of variable are these:

$$I = (P1)$$
$$O = (not \ P1) \ and \ (P2) \ and \ (P3)$$
$$I/O = (P1) \ and \ (P2) \ and \ (P3)$$
$$W = (not \ P1) \ and \ (P2) \ and \ (not \ P3)$$
$$B = (P1) \ and \ (P2) \ and \ (not \ P3)$$

In particular, the user must be careful not to change inadvertently the values of variables designated *I/O* or *B* between calls of the subroutine, whereas variables of designation *O,W* may freely be used for any purposes by the calling program. The user need not concern himself otherwise with *B* variables as they are generated by the subroutine itself.

(d) The last column of the Specifications list gives a brief description of the variable as it appears in the program.

STRUCTURE OF THE "NEXT" PROGRAMS

The six programs of NEX... type are alike in their bookkeeping relationships to the calling program. It was thought desirable for the subroutines to do as much of the bookkeeping as possible, and, to achieve that end, the following programming format has been observed:

(a) There is a subroutine variable MTC (mnemonic: "More To Come"). This variable is LOGICAL, and is named in the calling statements of the six "next" routines. When the subroutine returns to the main program, MTC will be set to either .TRUE. or .FALSE. If .TRUE., then the output which is being returned by the subroutine is not the last object in the collection of objects which is being searched (there are "More To Come"). If .FALSE., then current output is the last. Thus the calling program need only test MTC in order to determine if the search is complete. The subroutine itself carries the burden of knowing when the last object has been produced.

(b)   Whenever the user wishes to start a new search, whether this be the first request for a search, or whether, for any reason, he wants to restart the subroutine from the beginning of its list of objects, it is only necessary for the user to set MTC=.FALSE. himself prior to the call. The subroutines test MTC on entry, and if it is .FALSE., they reset themselves back to the first object corresponding to the current parameter values.

A typical use of a NEXT subroutine will look like this in the calling program:

```
      {Set parameters N,K,...}
      MTC=.FALSE.
10    CALL NEX...
      {Process output object}
      IF (MTC) GO TO 10
      . . .
```

STRUCTURE OF THE "RANDOM" PROGRAMS

The eight programs of random type listed above require a random number generator. A random number is a sample ξ drawn from a population which is uniformly distributed on the interval $0 < \xi < 1$. The preparation of such generators is discussed fully in several standard references. Also, on many computers, random number generators are built-in. The programs in this book expect that a compiled FUNCTION subprogram is available of the form

```
      FUNCTION RAND(I)
      . . .
      . . .
      RAND=...
      RETURN
      END
```

to be supplied by the user. We refrain from showing one here because such subprograms tend to be strongly machine-dependent. We follow the convention that each appearance of the letter ξ in a flow chart calls for the selection of a new random number.

ARRAYS AND SPECIFICATIONS

One of the main problems which must be confronted in the preparation of a collection of combinatorial subroutines is that of the organization of array storage. The choice of the correct array can often save considerable computing time, and in most of our applications this is very important. Yet the proliferation of large numbers of arrays may lead to insuperable storage problems when several subroutines are compiled together for some large application. We have been very conscious of these problems, and we have made certain policy decisions regarding the handling of arrays, which we now describe.

First, suppose a subroutine makes use of several arrays *in addition to* those which are of primary interest to the user. In FORTRAN, arrays which are not named in the calling statement must receive fixed dimensions in the subroutine itself, whereas the arrays which are named in the calling statement can have dimensions inherited from the main program.

Every such assignment of fixed dimension to an array whose actual length varies from one call of the subroutine to the next introduces a limitation on the capacity of the program. It is true that for different applications the user could change the DIMENSION statements to suit his needs, but at considerable inconvenience. This would lead to potentially frustrating problems for the user, and so (rightly or not!) we have almost invariably followed

Policy 1: The use of hidden arrays is avoided. More positively, every array used by a subroutine is named in its calling statement, unless the array is used only internally, and its dimension places no restriction on the subroutine parameters.

This policy has several corollaries, some pleasant and some not so. For example, two arrays of the same type which appear in two independent subroutines used for working storage only can be given the same name, thereby saving space.

As regards dimensioning, there are two main types of compilers. In one type, a dimension statement of the form

(∗) DIMENSION A(1), B(1, 1),...

appears in the subroutine and causes it to use the fixed dimensions assigned in the main program. In the second type, dimension statements like

(†) DIMENSION A(N), B(M,N), C(K),...

appear, where K,M,N,... are calling variables, and these may cause array space to be assigned according to the values of K,M,N,... at execution time. We have, quite arbitrarily, selected the second alternative here and followed it.

Policy 2: All arrays used by subroutines in this book carry dimension statements of the type (†) above.

One of the effects of Policy 1 is that calling statements are somewhat longer than they would be if hidden arrays were used. This is rarely troublesome since most subroutines have fewer than 6 variables in their parentheses, and only two subroutines have as many as 10 such variables. In order to achieve this economy we have been very aware of the need to avoid the use of unnecessary arrays, and we have not shrunk from doing even a small amount of extra computing toward that end.

An extreme example is Chapter 17, a subroutine whose sole raison d'etre is to avoid the use of an extra matrix array, and which, as a result, poses some very entertaining problems of programming and mathematics. Another small illustration is in Chapter 18 where a well-known search algorithm was not used, and a new one devised instead, just to avoid extra array storage. We state this, in summary, as

Policy 3: When confronted with the choice of a small amount of extra computation versus an extra array, save the array and do the computation.

We are under no illusions that these policies are ideal in all circumstances, but we do feel that they have been generally quite successful, and we should at this time take the opportunity to make the reader aware of them.

Part 1
Combinatorial Families

1

Next Subset of an n-Set (NEXSUB/LEXSUB)

As our introduction to combinatorial algorithms, we consider the question of generating all of the 2^n subsets of the set $\{1, 2, \ldots, n\}$. This will offer an opportunity to discuss the relationship between an algorithm and its proposed application. In fact, it is a truism that one should choose the method which best fits the problem, and we illustrate this by giving three different methods which are suited to different kinds of uses.

Let us imagine, then, a machine which, on request, displays a subset S of $\{1, 2, \ldots, n\}$. Following this, the user of the machine then does some calculation $C(S)$ with the set S. The user then asks for the next subset, etc., until all subsets have been processed. In this chapter we are concerned with the design of the subset-machine, under various hypotheses about the external calculation $C(S)$.

First we will suppose that there is nothing special about $C(S)$ at all, in which case the simplest possible design of the subset algorithm will be best.

Secondly, we will imagine that if S and S' are two consecutive sets produced by the algorithm, then S and S' are constrained to differ only by a *singleton*, for then the calculation $C(S)$ might be rapidly done by using the results of $C(S')$. An important example of such a calculation is in Chapter 23.

Finally, we will discuss the case where the external computation $C(S)$ is simplified if we can easily find the result of $C(S')$ for some set S' such that $S' \subset S$ and $|S'| = |S| - 1$.

(A) THE DIRECT APPROACH

To each subset $S \subseteq \{1, 2, \ldots, n\}$ we make correspond a binary number

$$m = a_1 + a_2 \cdot 2 + a_3 \cdot 2^2 + \cdots + a_n \cdot 2^{n-1}$$

by the relations

$$a_i = \begin{cases} 1 & \text{if } i \in S \\ 0 & \text{if } i \notin S \end{cases} \quad (i = 1, \ldots, n)$$

To go from a set S to its successor we simply replace m by $m + 1$ and read off the bits. Equivalently, we can operate directly on the bit-string $a_1, \ldots, a_n$ by simulating the operation $m \leftarrow m + 1$, and keeping track of the cardinality k, as follows:

(A) [*First entry*] $a_i \leftarrow 0$ $(i = 1, n)$; $k \leftarrow 0$; Exit.
(B) [*All later entries*] $i \leftarrow 1$.
(C) If $a_i = 0$, to **(D)**; $a_i \leftarrow 0$; $k \leftarrow k - 1$; $i \leftarrow i + 1$; to **(C)**.
(D) $a_i \leftarrow 1$; $k \leftarrow k + 1$; If $k = n$, final exit; Exit ∎

The labor per subset is measured by the index i in step **(D)**. We leave the reader to check that the average value of this index is

$$\sum_{i=1}^{n-1} i 2^{-i} \to 2 \quad (\text{as } n \to \infty).$$

The same algorithm can be stated in words: "to find the successor of a set S, insert the smallest element which is not in S, and delete all smaller elements from S."

(B) THE GRAY CODE

Now suppose each set S' is to differ from its immediate predecessor by the adjunction or deletion of a singleton. Here, for example, are the subsets of $\{1, 2, 3\}$ arranged in such a sequence:

$$\emptyset, \{1\}, \{1, 2\}, \{2\}, \{2, 3\}, \{1, 2, 3\}, \{1, 3\}, \{3\}.$$

Pictorially, consider the cube in 3-space whose vertices are the

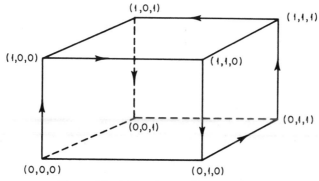

Figure 1.1 A Hamilton walk on the cube.

vectors of 0's and 1's. A sequence of sets such as the above corresponds to a walk along the edges of the cube, which starts at the origin and which visits each vertex exactly once. The list above, for instance, corresponds to the walk shown in Fig. 1.1.

A walk which visits every vertex of a graph exactly once is called a Hamilton walk on the graph. Hence a sequence of subsets of the desired type corresponds to a Hamilton walk on the n-cube, and our problem now is to describe such a walk algorithmically. We do this two ways, recursively and nonrecursively.

The recursive description of such a walk ("Gray code") is elegant. Let $\mathscr{L}_n$ denote, for each n, a Hamilton walk on the n-cube, and let $\overline{\mathscr{L}}_n$ denote the *reversed* walk (i.e., begin at the end and end at the origin). Then we define $\mathscr{L}_0$ to be the empty list and

(1) $$\mathscr{L}_n = \mathscr{L}_{n-1} \otimes 0, \overline{\mathscr{L}}_{n-1} \otimes 1 \qquad (n \geq 1)$$

The meaning of the notation is that we adjoin a symbol "0" to the right of each set in the list $\mathscr{L}_{n-1}$, then adjoin a symbol 1 to the right of each set in the reverse of the list $\mathscr{L}_{n-1}$, in order to obtain $\mathscr{L}_n$.

In this way we find, successively,

$\mathscr{L}_1$	$\mathscr{L}_2$	$\mathscr{L}_3$
0	00	000
1	10	100
	11	110
	01	010
		011
		111
		101
		001

etc.

We leave it to the reader to verify that the recursive description (1) does indeed provide a list of the desired type for each $n = 0$, 1, 2, . . . , and we proceed to discuss the nonrecursive construction of $\mathscr{L}_n$.

Suppose, generically, that we have arrived at a certain set S in the Gray code list $\mathscr{L}_n$ and that we wish to find the successor of S. In other words, we want to find the index j of the single coordinate a_j, which is to be changed in order to form the successor.

For example, the list $\mathscr{L}_4$ is as follows:

$$0000, \ 1000, \ 1100, \ 0100, \ 0110, \ 1110, \ 1010, \ 0010,$$
$$0011, \ 1011, \ 1111, \ 0111, \ 0101, \ 1101, \ 1001, \ 0001$$

The sequence of indices j of the changed coordinates is

$$1, \ 2, \ 1, \ 3, \ 1, \ 2, \ 1, \ 4, \ 1, \ 2, \ 1, \ 3, \ 1, \ 2, \ 1$$

The index j can be found from the set $S = \{a_1, \ . . . , a_n\}$ and its cardinality $k = |S|$ by the following

Rule of Succession If k is even, then $j = 1$; if k is odd, then j is the index of the coordinate which follows the first "1" bit of S.

The reader may wish to check this rule on the list $\mathscr{L}_4$ shown above.

It is easy to prove the correctness of the rule of succession directly from the recurrence (1) by induction. Indeed, if the rule does indeed form the lists $\mathscr{L}_1, \ . . . , \mathscr{L}_{n-1}$ and if the list $\mathscr{L}_i$ begins with $\varnothing$ and ends with $\{0, 0, 0, \ . . . , 0, 1\}$ for $1 \le i \le n - 1$, then from (1) we see that applying the rule clearly forms the first half of the list $\mathscr{L}_n$.

Further, one more application of the rule puts us at the beginning of $\overline{\mathscr{L}_{n-1}} \otimes 1$, the second half of $\mathscr{L}_n$. From there onward, the final "1" bit reverses the parity of the cardinality of each set S from what it was in $\mathscr{L}_{n-1}$. Since the rule never "sees" the final 1 bit, it produces, for each set S, exactly the index j that produced it from its predecessor in $\mathscr{L}_{n-1} \otimes 1$, i.e., which will produce its successor in the list $\overline{\mathscr{L}_{n-1}} \otimes 1$ ∎

We see that the rule can be applied if we know only the set S and one single bit of information more, namely the parity of $|S|$, in $O(1)$ time, on the average, per set. If we also want to recognize the last set in the list, when it is produced, in $O(1)$ time, we shall need $|S|$ itself and not just its parity.

This leads to the following algorithm.

ALGORITHM NEXSUB

[Generates the subsets of an n-set by the Gray code]

(A) [*First entry*] $a_i \leftarrow 0$ $(i = 1, n)$; $k \leftarrow 0$; Exit.
(B) [*Following entries*] $t \leftarrow \mod(k, 2)$; $j \leftarrow 1$; If $t \neq 0$, to (D).
(C) [*Change jth bit*] $a_j \leftarrow 1 - a_j$; $k \leftarrow k + 2a_j - 1$; If $k = a_n$, final exit; Exit.
(D) [*Find first 1 bit*] $j \leftarrow j + 1$; If $a_{j-1} = 1$, to (C); To (D) ■

(C) LEXICOGRAPHIC SEQUENCING

Our final arrangement of the subsets of $\{1, \ldots, n\}$ will be the lexicographic sequence.

Let us represent S by $(a_1, \ldots, a_k)$, a listing of its k members, in increasing order. Then we observe that $S < S' = (a_1', \ldots a_{k'}')$ in lexicographical order if there is $j \leq \min(k, k')$ such that $a_i = a_i'$ for $i < j$ and $a_j < a_j'$, or if $a_i = a_i'$ for $i \leq \min(k, k')$ and $k < k'$. This sequencing has an important property, which can be phrased in two ways.

(i)   All the sets S' obtained from a set $S = (a_1, \ldots, a_k)$ by adjoining to S only members greater than a_k form a contiguous sequence which immediately follows S.

(ii)   For any $S = (a_1, \ldots, a_k)$ the most recent set of i elements $(i < k)$ which precedes S in the sequencing is precisely the set $(a_1, \ldots, a_i)$.

Now back to the calculations $C(S)$, performed in lexicographical order. Set aside n pieces of storage (stack); when a $C(S)$ is performed, store its result in the kth piece of storage, where $k = |S|$. It is then easy to see that the information needed to facilitate this calculation is always found in the $(k - 1)$th piece of storage.

Some applications require the skipping of part of the sets in the lexicographical sequence, e.g., there may be a bound $k_{\max}$ on the cardinality. Or, for example, in calculating $C(S)$, it may be clear that the supersets of S are no longer desired. To cover these cases, algorithm LEXSUB has a parameter J, which causes a jump over all supersets following the input set, if $J =$ True. To use LEXSUB, the user must make his own initialization (by setting $k \leftarrow 0$); he may wish to stop at the last set ($a_1 = n$), or just after, when the null set is again produced ($k = 0$).

ALGORITHM LEXSUB

[Produces the successor of a subset of an *n*-set in lexicographical order. See text for initialization and termination.]

(A) [*Input n, k, J, (a_1, . . . , a_k): eliminate trivial cases; initialize*] If $k \neq 0$, to (B); if J = True, Exit; $s \leftarrow 0$; to (C).

(B) [*Test a_k and k for maximal values*] If $a_k = n$, to (E); $s \leftarrow a_k$; if J = True, to (D).

(C) $k \leftarrow k + 1$.

(D) $a_k \leftarrow s + 1$; Exit.

(E) [*Back up if $a_k = n$*] $k \leftarrow k - 1$; if $k = 0$, Exit; $s \leftarrow a_k$; to (D) ∎

SUBROUTINE SPECIFICATIONS (NEXSUB)

(1) *Name of subroutine:* NEXSUB.

(2) *Calling statement:* CALL NEXSUB(N,IN,MTC,NCARD,J).

(3) *Purpose of subroutine:* Generate subsets of $\{1, 2, . . . , N\}$, in an order specified by the Gray code.

(4) *Descriptions of variables in calling statement:*

Name	Type	I/O/W/B	Description
N	INTEGER	I	Number of elements in universe.
IN	INTEGER(N)	I/O	IN(I) = 1 if I is in output set; 0 if I is not in output set (I=1,N).
MTC	LOGICAL	I/O	TRUE. if current output set is not the last one; FALSE. if no more sets remain after current output.
NCARD	INTEGER	I/O	Cardinality of output set.
J	INTEGER	O	Index of coordinate changed to create current output set from previous (not available on first output).

(5) *Other routines which are called by this one:* None.

(6) *Number of* FORTRAN *instructions:* 19.

(7) *Remarks:* User supplies MTC=.FALSE. to call for a new sequence. The first output set is the empty set.

(8) To convert to a nonself-starting, memoryless cyclic production of successors, delete MTC from the instruction marked *, delete all instructions marked ** and add the instruction marked *** by removing the C in column 1.

```
      SUBROUTINE NEXSUB(N,IN,MTC,NCARD,J)              *
      INTEGER IN(N)
      LOGICAL MTC                                      **
      IF(MTC) GO TO 20                                 **
      DO 11 I=1,N                                      **
11    IN(I)=0                                          **
      NCARD=0                                          **
      MTC=.TRUE.                                       **
      RETURN                                           **
20    J=1
      IF(MOD(NCARD,2) .EQ. 0) GO TO 30
40    J=J+1
      IF(IN(J-1) .EQ. 0) GO TO 40
C     IF(J .GT. N) J=N                                 ***
30    IN(J)=1-IN(J)
      NCARD=NCARD+2*IN(J)-1
      MTC=NCARD .NE. IN(N)                             **
      RETURN
      END
```

SUBROUTINE SPECIFICATIONS (LEXSUB)

(1) *Name of subroutine:* LEXSUB.
(2) *Calling statement:* CALL LEXSUB(N,K,IN,JMP,NDIM).
(3) *Purpose of subroutine:* Generate subsets of $\{1, \ldots, N\}$ which succeed input set, in lexicographical order, with optional jumps over supersets.
(4) *Description of variables in calling statement:*

Name	Type	I/O/W/B	Description
N	INTEGER	I	Number of elements in universe.
NDIM	INTEGER	I	Maximal size of subset.
K	INTEGER	I/O	Cardinality of subset
IN	INTEGER(N)	I/O	IN(I), I=1,K is the Ith element of subset, listed in increasing order.
JMP	LOGICAL	I	To jump over supersets of input set, set JMP=.TRUE.; else set JMP=.FALSE.

(5) *Other routines which are called by this one:* None.
(6) *Number of* FORTRAN *instructions:* 16.
(7) *Remarks:* Subroutine has no memory or MTC ; it supplies the null set (K=0) after the last set in lexicographical order; this is also the starting point, which the user must supply (set K=0) .

```
       SUBROUTINE LEXSUB(N,K,IN,JMP,NDIM)
       DIMENSION IN(NDIM)
       LOGICAL JMP
10     IF(K .NE. 0) GO TO 40
20     IF(JMP) RETURN
30     IS=0
100    IF(.NOT. JMP) K=K+1
110    IN(K)=IS+1
       RETURN
40     IF(IN(K) .EQ. N) GO TO 50
80     IS=IN(K)
90     GO TO 100
50     K=K-1
60     IF(K .EQ. 0) RETURN
       IS=IN(K)
       GO TO 110
       END
```

SAMPLE OUTPUT (NEXSUB)

In the listing below are the 32 subsets of $\{1, 2, \ldots, 5\}$, one on each line, as produced by NEXSUB. On a line are IN(1),IN(2), ...,IN(5), followed by J, the index of the coordinate which was changed, and NCARD. Note how each line agrees with its predecessor except in the Jth entry.

0	0	0	0	0	0	0
1	0	0	0	0	1	1
1	1	0	0	0	2	2
0	1	0	0	0	1	1
0	1	1	0	0	3	2
1	1	1	0	0	1	3
1	0	1	0	0	2	2
0	0	1	0	0	1	1

0	0	1	1	0	4	2
1	0	1	1	0	1	3
1	1	1	1	0	2	4
0	1	1	1	0	1	3
0	1	0	1	0	3	2
1	1	0	1	0	1	3
1	0	0	1	0	2	2
0	0	0	1	0	1	1
0	0	0	1	1	5	2
1	0	0	1	1	1	3
1	1	0	1	1	2	4
0	1	0	1	1	1	3
0	1	1	1	1	3	4
1	1	1	1	1	1	5
1	0	1	1	1	2	4
0	0	1	1	1	1	3
0	0	1	0	1	4	2
1	0	1	0	1	1	3
1	1	1	0	1	2	4
0	1	1	0	1	1	3
0	1	0	0	1	3	2
1	1	0	0	1	1	3
1	0	0	0	1	2	2
0	0	0	0	1	1	1

SAMPLE OUTPUT (LEXSUB)

In the listing below are the 41 nonempty subsets of cardinality $\leq$ 3 of $\{1, 2, \ldots, 6\}$, as produced by calls to LEXSUB(6,K,IN,K.EQ.3,3)

```
1
1   2
1   2   3
1   2   4
1   2   5
1   2   6
1   3
1   3   4
1   3   5
1   3   6
```

```
1   4
1   4   5
1   4   6
1   5
1   5   6
1   6
2
2   3
2   3   4
2   3   5
2   3   6
2   4
2   4   5
2   4   6
2   5
2   5   6
2   6
3
3   4
3   4   5
3   4   6
3   5
3   5   6
3   6
4
4   5
4   5   6
4   6
5
5   6
6
```

2

Random Subset of an *n*-Set (RANSUB)

It is quite trivial to select a random subset of $\{1, 2, \ldots, n\}$: We flip a coin n times. If the ith toss is heads, then letter i belongs to the subset, otherwise it does not belong. If $a_i (i = 1, \ldots, n)$ is a random variable such that $a_i = 1$ or 0, depending on whether i belongs or does not belong to our set, then the algorithm is as follows:

ALGORITHM RANSUB

(A) $a_i \leftarrow \lfloor 2\xi \rfloor$ $(i = 1, n)$; Exit ■

The symbol $\lfloor x \rfloor$ denotes the largest integer $\leq x$.

As each of the elements of the set is chosen "in" or "out" with equal probability $\frac{1}{2}$, each of the 2^n sets is chosen with *uniform* probability 2^{-n}.

SUBROUTINE SPECIFICATIONS

(1) *Name of subroutine:* RANSUB.
(2) *Calling statement:* CALL RANSUB(N,A).

(3) *Purpose of subroutine:* Generate random subset of an *n*-set.
(4) *Descriptions of variables in calling statement:*

Name	Type	I/O/W/B	Description
N	INTEGER	I	Number of elements in set.
A	INTEGER(N)	O	A(I)=1 if I is in output set; 0 otherwise (I=1,N).

(5) *Other routines which are called by this one:* FUNCTION RAND(I) (random numbers).
(6) *Number of* FORTRAN *instructions:* 6.

```
      SUBROUTINE RANSUB(N,A)
      INTEGER A(N)
      DO 10  I=1,N
10    A(I)=2.*RAND(1)
      RETURN
      END
```

SAMPLE OUTPUT

The program RANSUB was called 1280 times with N=5. If all the 32 sets were chosen exactly the same number of times, the number of choices of each would be 40. It would be highly unlikely, however, that these exact numbers would be achieved. The following output shows the frequency with which each of the 32 subsets was selected, e.g., the empty set 43 times, etc. A measure of the likelihood of this distribution of frequencies is obtained from the so-called *chi-square test,* in which

$$\chi^2 = \sum_S \frac{(\phi(S) - 40)^2}{40}$$

Here $\phi(S)$ is the frequency of the subset S and the sum is over all 32 subsets. We find $\chi^2 = 21.35$. Statistical tables show that in 95% of such experiments, the observed value of χ^2 would lie between 17.5 and 48.2 if, indeed, all subsets were equally likely to be chosen.

```
0  0  0  0  0    43
1  0  0  0  0    33
0  1  0  0  0    41
1  1  0  0  0    53
0  0  1  0  0    38
1  0  1  0  0    35
0  1  1  0  0    40
1  1  1  0  0    37
0  0  0  1  0    39
1  0  0  1  0    44
0  1  0  1  0    42
1  1  0  1  0    41
0  0  1  1  0    44
1  0  1  1  0    38
0  1  1  1  0    47
1  1  1  1  0    39
0  0  0  0  1    47
1  0  0  0  1    39
0  1  0  0  1    32
1  1  0  0  1    34
0  0  1  0  1    30
1  0  1  0  1    48
0  1  1  0  1    37
1  1  1  0  1    43
0  0  0  1  1    38
1  0  0  1  1    44
0  1  0  1  1    35
1  1  0  1  1    37
0  0  1  1  1    34
1  0  1  1  1    44
0  1  1  1  1    37
1  1  1  1  1    47
```

CHI SQ IS 21.35 WITH 31 DEG FREEDOM

3

Next *k*-Subset of an *n*-Set
(NEXKSB/NXKSRD)

We consider here the combinations of n things taken k at a time. There are $\binom{n}{k}$ k-subsets of an n-set altogether, and in this chapter we give two different methods for generating them all, sequentially. In the first method we construct them in "alphabetical order," yielding a very simple algorithm. Following that, we describe a "revolving-door" method which generates each subset from its immediate predecessor by deleting some single element and adjoining some other single element. The lists of the ten 3-subsets of $\{1, 2, 3, 4, 5\}$ in these two orders are

$\{1, 2, 3\}$	$\{1, 2, 3\}$
$\{1, 2, 4\}$	$\{1, 3, 4\}$
$\{1, 2, 5\}$	$\{2, 3, 4\}$
$\{1, 3, 4\}$	$\{1, 2, 4\}$
$\{1, 3, 5\}$	$\{1, 4, 5\}$
$\{1, 4, 5\}$	$\{2, 4, 5\}$
$\{2, 3, 4\}$	$\{3, 4, 5\}$
$\{2, 3, 5\}$	$\{1, 3, 5\}$
$\{2, 4, 5\}$	$\{2, 3, 5\}$
$\{3, 4, 5\}$	$\{1, 2, 5\}$

In the lexicographic sequence, we obtain the successor of a given k-subset $\{a_1, \ldots, a_k\}$ as follows: search for the smallest h such that $a_{k+1-h} < n+1-h$, then increase a_{k+1-h} by 1, and set $a_j \leftarrow a_{j-1} + 1 (j = k+2-h, k)$. It is interesting to notice that the index h can be found *without searching*. Indeed, at each transition from a set to its successor, h increases by 1 unless $a_{k+1-h} < n-h$, in which case h is reset to 1 on the next transition.

ALGORITHM NEXKSB (LEXICOGRAPHIC)

(A) [*First entry*] $m \leftarrow 0$; $h \leftarrow k$; to **(D)**.
(B) [*Later entries*] If $m \geq n-h$, to **(C)**; $h \leftarrow 0$.
(C) $h \leftarrow h+1$; $m \leftarrow a_{k+1-h}$.
(D) For $j = 1, h$: $\{a_{k+j-h} \leftarrow m+j\}$; If $a_1 = n-k+1$, final exit; Exit ∎

It is not hard to measure the average amount of computational labor per subset generated. We claim that less than two units of labor are required, on the average. The index h measures the amount of labor per subset. For a fixed l, the number of k-subsets with $h = l+1$, i.e., with

$$a_k = n, \ a_{k-1} = n-1, \ \ldots, \ a_{k-l+1} = n-l+1 \quad \text{and} \quad a_{k-l} < n-l$$

is exactly

$$\binom{n-l-1}{k-l}$$

since $a_1, \ldots, a_{k-l}$ can be any $(k-l)$-subset of $\{1, 2, \ldots, n-l-1\}$. It follows that

$$\sum_{l=0}^{k} \binom{n-l-1}{k-l} = \binom{n}{k}$$

since every k-subset contributes exactly once to the left side, and furthermore, the average value of $h-1$ is (using the last sum also with k replaced by $k-1$)

$$\binom{n}{k}^{-1} \sum_{l=1}^{k} l \binom{n-l-1}{k-l}$$

$$= \binom{n}{k}^{-1} \left[k \sum_{l=0}^{k} \binom{n-k-1}{k-l} - \sum_{l=0}^{k-1} (k-l) \binom{n-l-1}{k-l} \right] \quad \text{(cont.)}$$

$$= \binom{n}{k}^{-1} \left[k \binom{n}{k} - \sum_{l=0}^{k-1} (n-k) \binom{n-l-1}{k-l-1} \right]$$

$$= k - \binom{n}{k}^{-1} (n-k) \binom{n}{k-1} = k - \frac{k(n-k)}{n-k+1} = \frac{k}{n-k+1}$$

Thus, if $k < (n/2)$, we do less than two units of labor per k-subset, on the average.

If we average h again over all 2^n subsets, we find that the average amount of labor per subset of an n-set is

$$2 - 2^{-n}$$

as claimed. A request for a subset is therefore quite inexpensive!

We turn now to the revolving-door (RD) algorithm. The motivation for this is similar to that in Chapter 1, namely, if we want to do a calculation for each subset, then if a subset differs only slightly from its predecessor, we may be able to save much of the calculation from the predecessor, thereby saving time. Since the cardinality k is fixed, to differ only slightly means that each subset is obtained by ejecting one element and adjoining another. An additional feature of RD is that the last k-subset on the list is next to the first one on the list in that one more "turn of the door" will return us to the beginning.

It is very easy to prove that such an algorithm exists for each n, $k (n = 1, 2, \ldots ; 0 \leq k \leq n)$. Indeed, let $A(m, l)$ denote a list of all of the l-subsets of $\{1, 2, \ldots, m\}$ arranged in RD order, beginning with ‡

$$\{1, 2, \ldots, l\}$$

and ending with

$$\{1, 2, \ldots, l-1, m\}$$

Then we have (the bar means reverse order)

(1) $A(n, k) = A(n-1, k), \overline{A(n-1, k-1)} \otimes \{n\}$

In other words, we construct $A(n, k)$ by first forming the list $A(n-1, k)$, and following it with the list $A(n-1, k-1)$ in reverse order with the singleton $\{n\}$ adjoined to each subset. It is then simple to check that if $A(n-1, k)$, and $A(n-1, k-1)$ are in RD order, then so is $A(n, k)$. It follows, by induction, that a list $A(n, k)$ exists for each n, k, starting with the inevitable $A(n, 0) = \varnothing$ for $n \geq 0$, and $A(n, n) = \{1, \ldots, n\}$ for $n > 0$.

For example, one easily derives

‡ There are other possibilities which, however, do not lead to much simpler algorithms.

(2) $$A(n, 1) = \{1\}, \{2\}, \ldots, \{n\}$$

and

(3) $$A(n, 2) = \{1, 2\}, \{2, 3\}, \{1, 3\}, \{3, 4\}, \{2, 4\}, \{1, 4\},$$
$$\ldots, \{1, n\}$$

To make the existence proof into an algorithm, we consider some samples. First, to find the successor of $C = \{4, 5, 7, 8\}$ in $A(9, 4)$, or in compressed notation ($S = $ successor, $P = $ predecessor), using (1)

$$S(C) = S\{4, 5, 7, 8\} \in A(9, 4) = A(8, 4), \overline{A(8, 3)} \otimes \{9\}$$

Note that $9 \notin C$, hence $C \in A(8, 4)$ and we have to look for

$$S(C) \in A(8, 4) = A(7, 4), \overline{A(7, 3)} \otimes \{8\}$$

Since $8 \in C$, C belongs to the second list, and because of the reversal of the listing, the problem is reduced to finding

$$P\{4, 5, 7\} \in A(7, 3) = A(6, 3), \overline{A(6, 2)} \otimes 7; \quad \text{add } \{8\}$$

As $7 \in \{4, 5, 7\}$, again the second list is needed, and we reverse again:

$$S\{4, 5\} \in A(6, 2) = A(5, 2), \overline{A(5, 1)} \otimes \{6\}; \quad \text{add } \{7, 8\}$$

Now no reversal as $6 \notin \{4, 5\}$

$$S\{4, 5\} \in A(5, 2) = A(4, 2), A(4, 1) \otimes \{5\}; \quad \text{add } \{7, 8\}$$

Another reversal, and we end up with

$$P\{4\} \in A(4, 1); \quad \text{add } \{5, 7, 8\}$$

Since $P\{4\} = \{3\}$ by (2), we have found

$$S\{4, 5, 7, 8\} = \{3, 5, 7, 8\}$$

Actually, the calculation could have proceeded much faster. The number of reversals for $\{4\}$ is equal to the number of elements in C larger than 4, that is 3, which is odd. Hence, we end up with $P\{4\} = \{3\}$, and add $\{4, 7, 8\}$.

If we want the successor of $\{4, 5, 7\}$ in $A(9, 3)$, however, a similar consideration would erroneously lead to asking for $S\{4\} \in A(4, 1)$, which does not exist. Backing up one step, however, we face the correct question

$$P\{4, 5\} \in A(5, 2) = A(4, 2), \overline{A(4, 1)} \otimes \{5\}; \quad \text{add } \{7\}$$

Now $\{4, 5\}$ belongs to $\overline{A(4, 1)} \otimes \{5\}$ and is its *first* element, so $P\{4, 5\}$ is the *last* element of $A(4, 2)$; that is $\{1, 4\}$. Hence,

$$S\{4, 5, 7\} = \{1, 4, 7\}$$

This second example was not resolved till we hit the separation between the two sublists of some list $A(n, k)$. Actually, the same should have happened in the first example as well, if we had not availed ourselves of (2). Recall that we were looking for $P\{4\} \in A(4, 1)$. Using (1) again, we have

$$P\{4\} \in A(4, 1) = A(3, 1), \overline{A(3, 0)} \otimes \{4\}$$

Now $\{4\}$ is the first (in fact, the only) member of $\overline{A(3, 0)} \otimes \{4\} = \{4\}$, so its predecessor is the last element of $A(3, 1)$; that is, $\{3\}$.

We now return to the general case. The calculation of $S\{a_1, \ldots, a_k\} \in A(n, k)$ leads, inductively, by (1), to the calculation of $S\{a_1, \ldots, a_j\}$ or $P\{a_1, \ldots, a_j\}$ in some $A(m, j)$. While in some initial step both $\{a_1, \ldots, a_k\}$ and its successor (or predecessor, as the case may be) belong to the same one of the segments $A(m - 1, j)$ or $A(m - 1, j - 1) \otimes \{m\}$, the decreasing of m at each step will force a first time when $\{a_1, \ldots, a_j\}$ and its successor (or predecessor) belong to different segments. When that happens we have two cases which are symbolized by

$$(4) \quad \{1, 2, \ldots, j - 1, m - 1\} \underset{P}{\overset{S}{\rightleftarrows}} \{1, 2, \ldots, j - 2, m - 1, m\}$$

$$\in A(m, j)$$

It is therefore imperative to find these values of m and j quickly. The examples suggest, rightly, that such a search should start from the left, i.e., with the small values of j and m. In fact, if we reduce j in the sets in (4) by one unit (the reader is invited to see that the same thing happens if j is further reduced) we would be looking for

$$P\{1, 2, \ldots, j - 1\} \in A(m, j)$$

or for

$$S\{1, 2, \ldots, j - 2, m - 1\} \in A(m - 1, j)$$

(the reduction of j increases the number of elements larger than the jth elements by one, and switches S and P), neither of which exist. This criterion of nonexistence provides a means of finding j and m in a left-to-right search: find the smallest j for which $P\{a_1, \ldots, a_j\}$ (or $S\{a_1, \ldots, a_j\}$, as required) exists.

Although the above discussion does not cover extreme values of j, the underlying concept has been explained. Further details are included in the flow chart and its description.

To estimate the labor involved in the revolving door algorithm we observe that if the set is of the form $\{1, \ldots, l, m, m'\}$, where $l + 1 < m < m'$, the algorithm loops l times; before it reaches boxes

70 or 150. In the $(l + 1)$th loop, if $k - l$ is even, we can now exit with a predecessor. (m' actually does not enter the consideration here, and need not exist.) If $k - l$ is odd, we can find a successor if $m' > m + 1$. (If there is no m', replace it by n; otherwise, one more loop is required.) In either case, l measures the amount of labor involved. The number of sets of the above form is $\binom{n-l-1}{k-l}$; this is exactly the same binomial coefficient as occurred in the computation of the labor for NEXKSB. Also the average value of l is computed by the same formula, so again no more than two units of labor is required when $k < n/2$.

The FORTRAN program for NXKSRD is supplied in two versions, one in which the subsets are produced sequentially, from the beginning, with an MTC indicating the end of the list; the second version simply supplies the *successor* to a subset supplied by the user, with the initial set following the last. The second version gives the user more flexibility, but requires also more user action. The same was not done for NEXKSB, because this program "remembers" the values of H and M from the previous call, and can therefore not be used to generate successors. The user may, however, convert the subroutine himself, e.g., by also making H and M I/O variables.

FLOW CHART NXKSRD

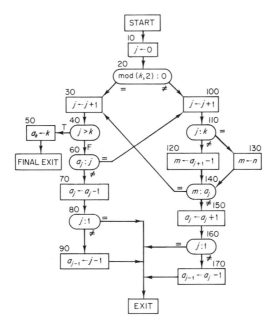

DESCRIPTION OF THE FLOW CHART

Box 10 Input n, k, $\{a_1, \ldots, a_k\}$. Initialization of j; the algorithm deals with subsets $\{a_1, \ldots, a_j\}$.

Box 20 If k is odd, we first look for a successor of $\{a_1\}$; otherwise, for a predecessor.

Box 30 The set $\{a_1, \ldots, a_j\}$ is either $\varnothing$ if $j = 0$ and we just entered the algorithm, or it is of the form $\{1, \ldots, j-1, m\}$, where $a_{j+1} = m + 1$, because Box 140 rejected the set as having no successor in $A(a_{j+1} - 1, j)$ [or in $A(n, k)$ if $j = k$]. Increment j and try for a predecessor of $\{1, \ldots, j-2, m, m+1\}$, or of $\{a_1\}$ if $j = 1$.

Box 40 There was no successor in $A(n, k)$, hence the set is $\{1, \ldots, k-1, n\}$.

Box 50 Restore original first input set.

Box 60 If $a_j = j$, the set is $\{1, \ldots, j\}$, and has no predecessor.

Box 70 The set is of the form $\{a_1\}$ with $a_1 > 1$ if $j = 1$, or of the form $\{1, \ldots, j-2, m, m+1\}$ if $j > 1$. Set $a_j \leftarrow a_j - 1$.

Box 80 If $j = 1$, the new set is finished. Old a_1 is removed, $a_1 - 1$ adjoined.

Box 90 If $j > 1$, set $a_{j-1} \leftarrow j - 1$. Now $m + 1$ is removed, while $j - 1$ is adjoined.

Box 100 The set is either $\varnothing$ if $j = 0$ and we just entered the algorithm, or it has no predecessor, and is therefore of the form $\{1, \ldots, j\}$. Increment j.

Box 110 We must look for a successor of $\{1, \ldots, j-1, a_j\}$ in $A(m, j)$ where m is to be determined.

Box 120 If $j < k$, $m = a_{j+1} - 1$.

Box 130 If $j = k$, $m = n$.

Box 140 If $a_j = m$, there is no successor of $\{1, \ldots, j-1, m\}$ in $A(m, j)$.

Box 150 The set is of the form $\{1, \ldots, j-1, a_j\}$. Set $a_j \leftarrow a_j + 1$.

Box 160 If $j = 1$, the set is complete; a_j has been removed, $a_j + 1$ has been adjoined.

Box 170 If $j > 1$, a_{j-1} needs to be changed. Now $j - 1$ has been removed and $a_j + 1$ adjoined.

SUBROUTINE SPECIFICATIONS (NEXKSB)

(1) *Name of subroutine:* NEXKSB.

(2) *Calling statement:* CALL NEXKSB(N,K,A,MTC).

(3) *Purpose of subroutine:* Next *k*-subset of an *n*-set, in lexicographic order.

(4) *Descriptions of variables in calling statement:*

Name	Type	I/O/W/B	Description
N	INTEGER	I	Number of elements in universe.
K	INTEGER	I	Number of elements in desired subset.
A	INTEGER(K)	I/O	A(I) is the Ith element of the output subset (I=1,K).
MTC	LOGICAL	I/O	To be set = .FALSE. before first call for a new sequence; = .TRUE. if current output is not the last subset; = .FALSE. if current output is the last.

(5) *Other routines which are called by this one:*   None.
(6) *Number of* FORTRAN *instructions:*   15.
(7) *Remarks:* $1 \leq A(1) < A(2) < \cdots \leq N$.

```
      SUBROUTINE NEXKSB(N,K,A,MTC)
      INTEGER A(K),H
      LOGICAL MTC
30    IF(MTC) GO TO 40
20    M2=0
      H=K
      GO TO 50
40    IF(M2.LT.N-H) H=0
      H=H+1
      M2=A(K+1-H)
50    DO 51 J=1,H
51    A(K+J-H)=M2+J
      MTC=A(1).NE.N-K+1
      RETURN
      END
```

SUBROUTINE SPECIFICATIONS (NXKSRD)

(1) *Name of subroutine:*   NXKSRD.
(2) *Calling statement:*   CALL NXKSRD(N,K,A,MTC,IN,OUT).
(3) *Purpose of subroutine:*   List *k*-subsets of an *n*-set, in RD order.
(4) *Descriptions of variables in calling statement:*

Name	Type	I/O/W/B	Description
N	INTEGER	I	Number of elements in universe.
K	INTEGER	I	Number of elements in desired subset.
A	INTEGER(K)	I/O	A(I) is the Ith element of the output subset (I=1,K).
MTC	LOGICAL	I/O	To be set =.FALSE. before first call for a new sequence; =.TRUE. if current output is not the last; =.FALSE. if no more subsets remain after current one.
IN	INTEGER	O	Element of output set which was not in input set.
OUT	INTEGER	O	Element of input set which is not in output set.

(5) *Other routines which are called by this one:*   None.
(6) *Number of* FORTRAN *instructions:*   34.

(7) *Remarks:* The program is converted to a no-MTC, no-memory version by deleting MTC from the instruction labeled *, by deleting instructions ** , and by inserting *** (remove the C from column 1). The program then simply calculates the successor of any input set, and starts again from the beginning after the last set.

```
        SUBROUTINE NXKSRD(N,K,A,MTC,IN,OUT)           *
        INTEGER A(K), OUT
        LOGICAL MTC                                   **
        IF(MTC) GO TO 10                              **
        DO 1 I=1,K                                    **
1       A(I)=I                                        **
        MTC=K .NE. N                                  **
        RETURN                                        **
10      J=0
20      IF(MOD(K,2) .NE. 0) GO TO 100
30      J=J+1
C 40    IF(J .LE. K) GO TO 60                         ***
C 50    A(K)=K                                        ***
C       IN=K                                          ***
C       OUT=N                                         ***
C       RETURN                                        ***
60      IF(A(J) .EQ. J) GO TO 100
70      OUT=A(J)
        IN=OUT-1
        A(J)=IN
80      IF(J .EQ. 1) GO TO 200
90      IN=J-1
        A(J-1)=IN
        GO TO 200
100     J=J+1
130     M=N
110     IF(J .LT. K) M=A(J+1)-1
140     IF(M .EQ. A(J)) GO TO 30
150     IN=A(J)+1
        A(J)=IN
        OUT=IN-1
        IF(J .EQ. 1) GO TO 200
        A(J-1)=OUT
        OUT=J-1
200     IF(K .EQ. 1) GO TO 201                        **
        MTC=A(K-1) .EQ. K-1                           **
```

```
201    MTC=(.NOT.MTC) .OR. A(K) .NE. N        **
       RETURN                                 **
C200   RETURN                                 ***
       END
```

SAMPLE OUTPUT (NEXKSB)

The program NEXKSB (in lexicographic order) was called, repeatedly, with N=7, K=4, until termination. The 35 output vectors A(1),A(2),A(3),A(4) are now shown.

```
1   2   3   4
1   2   3   5
1   2   3   6
1   2   3   7
1   2   4   5
1   2   4   6
1   2   4   7
1   2   5   6
1   2   5   7
1   2   6   7
1   3   4   5
1   3   4   6
1   3   4   7
1   3   5   6
1   3   5   7
1   3   6   7
1   4   5   6
1   4   5   7
1   4   6   7
1   5   6   7
2   3   4   5
2   3   4   6
2   3   4   7
2   3   5   6
2   3   5   7
2   3   6   7
2   4   5   6
2   4   5   7
2   4   6   7
2   5   6   7
3   4   5   6
```

```
3  4  5  7
3  4  6  7
3  5  6  7
4  5  6  7
```

SAMPLE OUTPUT (NXKSRD)

Below there appear the 4-subsets of $\{1, \ldots, 7\}$ and the 5-subsets of $\{1, \ldots, 8\}$ as output by the revolving-door subroutine. On each line, we show first the elements of the set, then the two elements IN, OUT which have just been exchanged.

```
1  2  3  4     0  0
1  2  4  5     5  3
2  3  4  5     3  1
1  3  4  5     1  2
1  2  3  5     2  4
1  2  5  6     6  3
2  3  5  6     3  1
1  3  5  6     1  2
3  4  5  6     4  1
2  4  5  6     2  3
1  4  5  6     1  2
1  2  4  6     2  5
2  3  4  6     3  1
1  3  4  6     1  2
1  2  3  6     2  4
1  2  6  7     7  3
2  3  6  7     3  1
1  3  6  7     1  2
3  4  6  7     4  1
2  4  6  7     2  3
1  4  6  7     1  2
4  5  6  7     5  1
3  5  6  7     3  4
2  5  6  7     2  3
1  5  6  7     1  2
1  2  5  7     2  6
2  3  5  7     3  1
1  3  5  7     1  2
3  4  5  7     4  1
```

2	4	5	7		2	3
1	4	5	7		1	2
1	2	4	7		2	5
2	3	4	7		3	1
1	3	4	7		1	2
1	2	3	7		2	4

1	2	3	4	5		0	0
1	2	3	5	6		6	4
1	3	4	5	6		4	2
2	3	4	5	6		2	1
1	2	4	5	6		1	3
1	2	3	4	6		3	5
1	2	3	6	7		7	4
1	3	4	6	7		4	2
2	3	4	6	7		2	1
1	2	4	6	7		1	3
1	4	5	6	7		5	2
2	4	5	6	7		2	1
3	4	5	6	7		3	2
1	3	5	6	7		1	4
2	3	5	6	7		2	1
1	2	5	6	7		1	3
1	2	3	5	7		3	6
1	3	4	5	7		4	2
2	3	4	5	7		2	1
1	2	4	5	7		1	3
1	2	3	4	7		3	5
1	2	3	7	8		8	4
1	3	4	7	8		4	2
2	3	4	7	8		2	1
1	2	4	7	8		1	3
1	4	5	7	8		5	2
2	4	5	7	8		2	1
3	4	5	7	8		3	2
1	3	5	7	8		1	4
2	3	5	7	8		2	1
1	2	5	7	8		1	3
1	5	6	7	8		6	2
2	5	6	7	8		2	1
3	5	6	7	8		3	2

```
4  5  6  7  8     4  3
1  4  6  7  8     1  5
2  4  6  7  8     2  1
3  4  6  7  8     3  2
1  3  6  7  8     1  4
2  3  6  7  8     2  1
1  2  6  7  8     1  3
1  2  3  6  8     3  7
1  3  4  6  8     4  2
2  3  4  6  8     2  1
1  2  4  6  8     1  3
1  4  5  6  8     5  2
2  4  5  6  8     2  1
3  4  5  6  8     3  2
1  3  5  6  8     1  4
2  3  5  6  8     2  1
1  2  5  6  8     1  3
1  2  3  5  8     3  6
1  3  4  5  8     4  2
2  3  4  5  8     2  1
1  2  4  5  8     1  3
1  2  3  4  8     3  5
```

4

Random k-Subset of an n-Set (RANKSB)

Suppose that integers k, n are given, $1 \leqq k \leqq n$, and we want to select k distinct elements $a_1, \ldots, a_k$ from $\{1, 2, \ldots, n\}$, at random. This innocent-sounding question in fact poses some substantive algorithmic problems. The output set is to contain k words, and so it seems reasonable to impose the requirement that

(i)   *no more than k words of array storage should be used by the algorithm, each word to hold an integer between 0 and n.*

Next, it seems that rather little labor should be needed beyond the selection of k integers and perhaps some removal of duplicates, and so our second condition is

(ii)   *the average labor required should be $O(k)$.*

Furthermore, in some applications (see, e.g., Chapter 6) though not in all it is helpful if the elements of the output set are presented in ascending order, and so we ask

(iii)   *for the output set we have $1 \leqq a_1 < a_2 < \cdots < a_k \leqq n$.*

The reader may wish, before reading on, to try to design a method which satisfies (i)–(iii). The main problem is that we select, one at a time, integers at random between 1 and n, and we need to know if the integer just chosen has been chosen before. If so, the new integer is discarded, otherwise it is kept.

But how shall we discover if the latest integer is "new"? If we examine all integers so far chosen, we end with $O(k^2)$ labor, in violation of (ii). If we arrange the integers in a linked tree, the labor drops to $O(k^{3/2})$, still in violation of (ii) and the links force a violation of (i) also. If we keep an array whose ith entry tells us whether i has been chosen, then (i) and (iii) will be violated.

Our algorithm, which meets all three requirements, is, in broad outline, to divide the range $[1, n]$ into k subintervals ("bins") $R_l(l = 1, \ldots, k)$ of approximately equal sizes, and choose the cardinalities $|B_l|$ of the sets B_l of elements to be chosen from each bin R_l. The $|B_l|$ have a multinomial distribution; we determine them by a rejection method which simulates the choosing and recording of the members of the B_l. When this step is finished, we now must choose the actual members of the B_l from the interval R_l, again uniformly at random. This problem is in essence the same as the one we originally faced. Now, however, the $|B_l|$ are very small, and we can afford to use a direct method which is quadratic in $|B_l|$.

First, let us analyze the number of random choices of an integer which must be made in order to obtain k distinct integers. It is well known [Kn1, Vol. II, p. 470] that the expected number of independent random drawings from $\{1, 2, \ldots, n\}$, which must be made in order to obtain k distinct samples, is

$$(1) \qquad \Delta = n \left\{ \frac{1}{n - k + 1} + \cdots + \frac{1}{n} \right\}$$

If $k = n$, for example, we need $\Delta \sim n \log n$ such drawings, on the average. If $k/n \leqq \theta < 1$, then the number of such drawings needed is

$$\Delta \lesssim \left(\frac{1}{\theta} \log \frac{1}{1 - \theta} \right) k$$

The first component of the labor required to execute our algorithm, namely, the work needed to select k distinct integers from $\{1, 2, \ldots, n\}$ by independent random samples, is therefore $O(k)$, provided $k \leqq n/2$, say, and is $O(k \log k)$ in any case.

Next, we plan to do the following.

(a)   Divide the range $[1, n]$ into the k subranges

$$R_l = \left\{ m \middle| \left\lfloor \frac{(l - 1)n}{k} \right\rfloor < m \leqq \left\lfloor \frac{ln}{k} \right\rfloor \right\} \qquad (l = 1, \ldots, k)$$

The random k-set to be chosen will consist of members B_l of the bins

R_l. As the number of sets B_l equals k, they will contain very few elements; some will be empty.

(b)   First determine the cardinalities $|B_l|$, without worrying about exactly which elements of R_l will be members of B_l. In doing this, the k storage locations a_l can be used, one for each B_l. We draw a random number x in the range $[1, n]$, determine the R_l to which it belongs, by

(2) $$l = 1 + \lfloor (xk - 1)/n \rfloor$$

and accept or reject the x depending on whether it "duplicates" an element already accepted. Suppose m members of R_l have already been accepted, while the total number of members of R_l is q, then the probability that x is rejected, is m/q. With this in mind, we may simply reject x with probability m/q, without ever checking x against any element that has been accepted! In fact, we only maintain a *count* of the elements that have been "accepted." All this (and a bit more) is accomplished as follows. Initially, we store in a_l the number $\lfloor (l-1)n/k \rfloor$; this is one unit less than the smallest element in R_l. When an x has been chosen, and R_l is determined by (2), we accept x if $x > a_l$, reject if $x \le a_l$. If x is "accepted," we increase a_l by one unit—but drop x (!).

(c)   When k such x have been accepted, we scan the $a_l (l = 1, \ldots , k)$ and move those a_l that no longer have their initial values (they represent the nonempty B_l) to the leftmost positions of the array, say $(a_1, \ldots , a_p)$. Next, for $j = p, p - 1, \ldots , 1$ we reserve space for each of the B_l, starting from the right. We determine by (2) the segment R_l to which a_j belongs, and determine the cardinality of B_l. Now we reserve $|B_l|$ spaces for B_l in $(a_1, \ldots , a_k)$, starting from the right, store the value of l in the rightmost of these spaces; the others are zeroed out.

(d)   Again, scanning from the right, we now place random elements of R_l into the space reserved for B_l, listing them in order. Duplications are avoided by choosing, at random, $m' = 1 + \lfloor \xi m \rfloor$, where m is the number of members of R_l not yet chosen, and x is the m'th member of this list of unchosen elements.

ALGORITHM RANKSB

[Input: n, k; output: random subset $(a_1, \ldots , a_k)$ of $\{1, \ldots , n\}$, listed in increasing order. Time: linear in k for $k \le n/2$; no auxiliary array storage.]

(A) [*Initialize a_i to "zero" point for bin R_i*] $a_i \leftarrow \lfloor (i-1)n/k \rfloor$
$(i = 1, k); c \leftarrow k$.

(B) [*Choose random x; determine range R_l; accept or reject*} $x \leftarrow 1 + \lfloor \xi n \rfloor$; $l \leftarrow 1 + \lfloor (xk-1)/n \rfloor$; if $x \le a_l$, to **(B)**;
$a_l \leftarrow a_l + 1$; $c \leftarrow c - 1$; if $c > 0$, to **(B)**; $i \leftarrow 0$; $p \leftarrow 0$; $s \leftarrow k$.

(C) [*Move a_i of nonempty bins to the left*] $i \leftarrow i + 1$; if $i > k$, to
(D); if $a_i = \lfloor (i-1)n/k \rfloor$, $a_i \leftarrow 0$ and to **(C)**; $p \leftarrow p + 1$; $m \leftarrow a_i$;
$a_i \leftarrow 0$; $a_p \leftarrow m$; to **(C)**.

(D) [*Determine l, set up space for B_l*] $l \leftarrow 1 + \lfloor (a_p k - 1)/n \rfloor$;
$\Delta s \leftarrow a_p - \lfloor (l-1)n/k \rfloor$; $a_p \leftarrow 0$; $a_s \leftarrow l$; $s \leftarrow s - \Delta s$; $p \leftarrow p - 1$;
if $p > 0$, to **(D)**; $l \leftarrow k$.

(E) [*If $a_l \ne 0$, a new bin is to be processed*] if $a_l = 0$, to **(F)**; $r \leftarrow l$;
$m_0 \leftarrow 1 + \lfloor (a_l - 1)n/k \rfloor$; $m \leftarrow \lfloor a_l n/k \rfloor - m_0 + 1$.

(F) [*Choose a random x*] $x \leftarrow m_0 + \lfloor \xi m \rfloor$; $i \leftarrow l$.

(G) [*Check x against previously entered elements in bin; increment
x as it jumps over elements $\le x$*] $i \leftarrow i + 1$; if $i > r$, to **(H)**; if
$x < a_i$, to **(H)**; $a_{i-1} \leftarrow a_i$; $x \leftarrow x + 1$; to **(G)**.

(H) [*Insert x; exit if last l*] $a_{i-1} \leftarrow x$; $m \leftarrow m - 1$; $l \leftarrow l - 1$; if
$l = 0$, Exit; to **(E)** ∎

A labor estimate is obtained as follows. All operations through **(D)**
are linear in k (if $k \le n/2$). Loop **(G)** resides inside loop **(E)** to **(H)**;
the labor for B_l is $O(|B_l|^2)$. To obtain an average labor, we must estimate the latter. We do so for large n; if n is not large compared to k,
then the larger values of m_i will become even less likely, due to a
higher probability of rejection.

The probability of getting subsets $B_1, \ldots, B_k$ containing
$m_1, \ldots, m_k$ numbers ($\Sigma m_i = k$) is

$$k^{-k} \binom{k}{m_1, \ldots, m_k}$$

The labor for each subset B_i is proportional to m_i^2. Hence, we calculate the total average labor $\mathscr{L}$:

$$\mathscr{L} = \Sigma k^{-k} \binom{k}{m_1, \ldots, m_k} \Sigma m_i^2$$

Because of symmetry we may replace Σm_i^2 by km_1^2;

$$\mathscr{L} = \Sigma k^{-k+1} \binom{k}{m_1, \ldots, m_k} m_1^2$$

The typical term in this sum is the coefficient of $x_1^{m_1} \cdots x_k^{m_k}$ in

$$k^{-k+1} \left(x_1 \frac{\partial}{\partial x_1} \right)^2 (x_1 + \cdots + x_k)^k$$

Calculating the latter and setting $x_1 = \cdots = x_k = 1$ we get $\mathscr{L} = 2k - 1$. Hence, the average labor is linear in k, for $k \leq n/2$, and $O(k \log k)$ for all k.

The required labor could have been held to $O(k)$ uniformly for $1 \leq k \leq n$ by using a different method when $k \geq n/2$. The following algorithm, for example, uses just k memory locations, operates in $O(k)$ steps when $n/2 \leq k \leq n$, and it also produces sorted output.

ALGORITHM RKS2

(A) $c_1 \leftarrow k$; $c_2 \leftarrow n$; $k_0 \leftarrow 0$; $i \leftarrow 0$.
(B) $i \leftarrow i + 1$; If $\xi > c_1/c_2$, to (C); $c_1 \leftarrow c_1 - 1$; $k_0 \leftarrow k_0 + 1$; $a_{k_0} \leftarrow i$; If $c_1 \leq 0$, Exit.
(C) $c_2 \leftarrow c_2 - 1$; To (B) ∎

In the interest of programming simplicity, we have chosen not to include this algorithm in our program.

SUBROUTINE SPECIFICATIONS

(1) *Name of subroutine:* RANKSB.
(2) *Calling statement:* CALL RANKSB(N,K,A).
(3) *Purpose of subroutine:* Choose a random k-subset of {1, 2, . . . , n}.
(4) *Description of variables in calling statement:*

Name	Type	I/O/W/B	Description
N	INTEGER	I	Number of elements in universe.
K	INTEGER	I	Number of elements in desired subset.
A	INTEGER(K)	O	A(I) is the Ith element of the output subset (I=1,K).

(5) *Other routines which are called by this one:* Random number generator FUNCTION RAND(I).
(6) *Number of* FORTRAN *instructions:* 46.
(7) *Remarks:* A(1),...,A(K) are in sorted sequence.

```
SUBROUTINE RANKSB(N,K,A)
INTEGER A(K),X,R,DS,P,S,C
C=K
```

```
      DO 1 I=1,K
1     A(I)=(I-1)*N/K
10    X=1+N*RAND(1)
      L=1+(X*K-1)/N
      IF(X .LE. A(L)) GO TO 10
      A(L)=A(L)+1
      C=C-1
      IF(C .NE. 0) GO TO 10
      P=0
      S=K
      DO 20 I=1,K
      M=A(I)
      A(I)=0
      IF(M .EQ. (I-1)*N/K) GO TO 20
      P=P+1
      A(P)=M
20    CONTINUE
30    L=1+(A(P)*K-1)/N
      DS=A(P)-(L-1)*N/K
      A(P)=0
      A(S)=L
      S=S-DS
      P=P-1
      IF(P .GT. 0) GO TO 30
      L=K
40    IF(A(L).EQ.0) GO TO 50
      R=L
      MO=1+(A(L)-1)*N/K
      M=A(L)*N/K-MO+1
50    X=MO+M*RAND(1)
      I=L
60    I=I+1
      IF(I .LE. R) GO TO 80
70    A(I-1)=X
      M=M-1
      L=L-1
      IF(L .EQ. 0) RETURN
      GO TO 40
80    IF(X .LT. A(I)) GO TO 70
      X=X+1
      A(I-1)=A(I)
      GO TO 60
      END
```

SAMPLE INTERMEDIATE RESULT

The program RANKSB was called with $n = 63$, $k = 10$. The vector **A** has been reproduced, after the completion of the indicated step of the algorithm.

(A)	0	6	12	18	25	31	37	44	50	56
(B)	1	6	12	18	27	33	39	44	50	59
(C)	1	27	33	39	59	0	0	0	0	0
(D)	1	0	5	0	6	0	7	0	0	10
(F)	$x = 61$									
(H)	1	0	5	0	6	0	7	0	0	61
(F)	$x = 57$									
(H)	1	0	5	0	6	0	7	0	57	61
(F)	$x = 58$									
(H)	1	0	5	0	6	0	7	57	59	61

The next two values of x were 38, 43

(H)	1	0	5	0	6	38	44	57	59	61

SAMPLE OUTPUT

The program RANKSB was called 200 times with $n = 5$, $k = 3$. The frequencies with which each of the ten subsets were obtained are shown below. Thus $\{1, 2, 3\}$ occurred 18 times, etc. The value of χ^2 is 7.24 with 9 degrees of freedom. In 95% of such experiments, the observed value of χ^2 would lie between 2.6 and 19.6 if all 3-subsets were equally likely to occur.

1	2	3	18
1	2	4	23
1	2	5	18
1	3	4	20
1	3	5	16
1	4	5	19
2	3	4	21
2	3	5	19
2	4	5	18
3	4	5	28

5

Next Composition of n into k Parts (NEXCOM)

Let n and k be fixed positive integers. By a *composition* of n into k parts, we mean a representation of the form

$$(1) \qquad n = r_1 + r_2 + \cdots + r_k$$

in which $r_i \geqq 0$ $(i = 1, k)$ and the order of the summands is important.

For example, there are exactly 28 compositions of 6 into 3 parts, namely,

$$
\begin{aligned}
6 &= 6 + 0 + 0 = 0 + 6 + 0 = 0 + 0 + 6 = 1 + 2 + 3 \\
&= 5 + 1 + 0 = 5 + 0 + 1 = 1 + 5 + 0 = 2 + 1 + 3 \\
&= 1 + 0 + 5 = 0 + 1 + 5 = 0 + 5 + 1 = 2 + 2 + 2 \\
&= 4 + 2 + 0 = 4 + 0 + 2 = 0 + 4 + 2 = 2 + 4 + 0 \\
&= 2 + 0 + 4 = 0 + 2 + 4 = 4 + 1 + 1 = 1 + 4 + 1 \\
&= 1 + 1 + 4 = 3 + 3 + 0 = 3 + 0 + 3 = 0 + 3 + 3 \\
&= 3 + 2 + 1 = 3 + 1 + 2 = 1 + 3 + 2 = 2 + 3 + 1
\end{aligned}
$$

We now derive a formula for $J(n, k)$, the number of compositions of n into k parts. The derivation will show, also, how to construct a simple algorithm for generating all of them.

Suppose that n indistinguishable balls are to be arranged in k labeled cells. There are evidently exactly $J(n, k)$ ways to do the arranging because, if we have such an arrangement, let r_i be the number of balls in the ith cell for $i = 1, k$. Then we have a composition (1), and the converse is also true. Hence, we can find $J(n, k)$ if we can count these arrangements of n balls in k cells.

Let $n + k + 1$ spaces be marked on a sheet of paper, and suppose that in the first space and the last space we mark a vertical bar, as

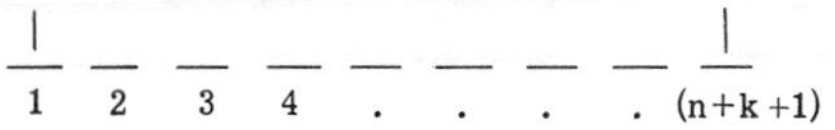

Figure 5.1

shown in Fig. 5.1. In the remaining $n + k - 1$ spaces, distribute the n balls with no more than one ball occupying any space. There are obviously

$$\binom{n + k - 1}{n}$$

ways of doing this. In each of the other $k - 1$ spaces which remain, place a vertical bar. We now have a pattern like the one shown in Fig. 5.2.

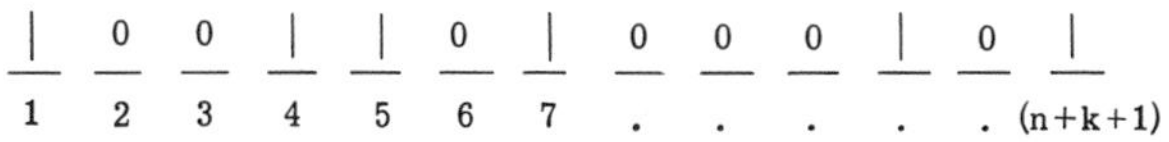

Figure 5.2

Now we think of the vertical bars as representing cell boundaries. Hence, in Fig. 5.2 there are 5 cells containing, respectively, 2, 0, 1, 3, 1 balls. It is now clear that there are precisely

(2) $$J(n, k) = \binom{n + k - 1}{n}$$

compositions of n into k parts. For example, the

$$\binom{6 + 3 - 1}{6} = \binom{8}{6} = 28$$

compositions of 6 into 3 parts have been listed above.

Another proof of the same result can be given, which, while it does not help with the design of an algorithm, shows an important area of applications. Indeed, suppose we are given a number of power series and we want to multiply them together. How can we calculate the coefficients of the product series? For instance, if

$$\left(\sum a_i x^i\right)\left(\sum b_j x^j\right)\left(\sum c_k x^k\right) = \sum d_m x^m$$

how can we express d_m in terms of a_i, b_j, c_k? Clearly,

$$(3) \qquad d_m = \sum_{i+j+k=m} a_i b_j c_k$$

On the right side of (3) there is a term corresponding to each composition of m into three parts.

Consider the power series $f(x) = 1 + x + x^2 + \cdots$. If we raise it to the kth power, we get

$$(4) \qquad f(x)^k = \sum_{r_1=0}^{\infty} \sum_{r_2=0}^{\infty} \cdots \sum_{r_k=0}^{\infty} x^{r_1+r_2+\cdots+r_k}$$

Collecting terms with equal exponents, we see that x^m appears exactly as often as there are compositions of m; hence,

$$(5) \qquad f(x)^k = \sum_{n=0}^{\infty} J(n, k) x^n$$

On the other hand, $f(x)^k = (1 - x)^{-k}$ but by Taylor's theorem,

$$(6) \qquad \frac{1}{(1-x)^k} = \sum \binom{n+k-1}{n} x^n$$

and comparison of (5) and (6) yields (2) again.

The algorithm for generating compositions of n into k parts sequentially is suggested by our first proof, above, of the relation (2). What we must do in order to generate all of the compositions of n into k parts is to generate all of the $(k-1)$-subsets of $n+k-1$ objects and to interpret each such subset as the set of locations of the interior vertical bars in Fig. 5.2. From the bar locations we can, by subtraction or otherwise, determine the number of balls between each consecutive pair of bars and thereby determine the composition which corresponds to the given subset.

Instead of generating the subsets and from each subset computing the composition, we can do both together by going back to our lexicographic algorithm NEXKSB and translating it into a direct algorithm for compositions. Recall that in that algorithm, if

$$(7) \qquad \{a_1, a_2, \ldots, a_{k-1}\}$$

is a $(k-1)$-subset, we go to the next one by finding the smallest h for which

(8) $a_{k-1} = n, a_{k-2} = n - 1, \ldots, a_{k-h} = n - h + 1; a_{k-h-1} < n - h$

We then increase a_{k-h-1} by 1 and set each succeeding a_{r+1} equal to one more than its predecessor a_r $(r = k - h - 1, \ldots, k - 2)$.

In terms of the composition associated with the subset (7)

(9) $$n = r_1 + r_2 + \cdots + r_k$$

the relations (8) imply that $r_k = r_{k-1} = \cdots = r_{k+1-h} = 0$ and $r_{k-h} > 0$. The act of increasing a_{k-h-1} by 1 and setting each following a_r equal to one more than its predecessor, will (a) increase r_{k-h-1} by 1, (b) set $r_k = r_{k-h} - 1$, and (c) set $r_{k-h} = 0$. The reader can easily follow this by watching what happens to the moving vertical bars in Fig. 5.2 and noticing that the end bars remain fixed.

The language of subsets can therefore be removed, and the entire algorithm can be stated directly in terms of compositions. It is also convenient to search for the first nonzero part of the composition starting from the left rather than from the right-hand side.

When all of this is done, what remains is to search the last composition $r_1, \ldots, r_k$ to find the first nonzero part r_h. We then put $t \leftarrow r_h$, $r_h \leftarrow 0$, $r_1 \leftarrow t - 1$ and $r_{h+1} \leftarrow r_{h+1} + 1$. It is important to notice, however, that the same line of measuring as we applied in Chapter 3 also works here, and we do not actually have to *search* for the first nonzero part.

Indeed, if the first nonzero part of the previous composition was > 1, then we will find $h = 1$ on the next composition, while if that first nonzero part was $= 1$, we will have $h \leftarrow h + 1$ on the next composition. The complete formal algorithm follows.

ALGORITHM NEXCOM

(A) [*First entry*] $r_1 \leftarrow n; t \leftarrow n; h \leftarrow 0; r_i \leftarrow 0$ $(i = 2, k)$; Go to (D).
(B) [*Later entries*] If $t = 1$, go to (C); $h \leftarrow 0$;
(C) $h \leftarrow h + 1; t \leftarrow r_h; r_h \leftarrow 0; r_1 \leftarrow t - 1; r_{h+1} \leftarrow r_{h+1} + 1.$
(D) If $r_k = n$ final exit; Exit ∎

SUBROUTINE SPECIFICATIONS

(1) *Name of subroutine:* NEXCOM.
(2) *Calling statement:* CALL NEXCOM(N,K,R,MTC).

(3) *Purpose of subroutine:* Next composition of n into k parts.

(4) *Descriptions of variables in calling statement:*

Name	Type	I/O/W/B	Description
N	INTEGER	I	Number whose compositions are desired.
K	INTEGER	I	Number of parts of desired composition.
R	INTEGER(K)	I/O	R(I) is the Ith part of the output composition (I=1,K).
MTC	LOGICAL	I/O	=.TRUE. if this is not the last composition; =.FALSE. if the current output is the last.

(5) *Other routines which are called by this one:* None.

(6) *Number of* FORTRAN *instructions:* 20.

```
      SUBROUTINE NEXCOM (N,K,R,MTC)
      INTEGER R(K),T,H
      LOGICAL MTC
10    IF(MTC) GO TO 20
      R(1)=N
      T=N
      H=0
      IF(K.EQ.1) GO TO 15
      DO 11 I=2,K
11    R(I)=0
15    MTC=R(K).NE.N
      RETURN
20    IF(T.GT.1) H=0
30    H=H+1
      T=R(H)
      R(H)=0
      R(1)=T-1
      R(H+1)=R(H+1)+1
      GO TO 15
      END
```

SAMPLE OUTPUT

The program NEXCOM was called, repeatedly, with N=6, K=3, until termination. The 28 output vectors R(1),R(2),R(3) are shown on the next page.

6	0	0
5	1	0
4	2	0
3	3	0
2	4	0
1	5	0
0	6	0
5	0	1
4	1	1
3	2	1
2	3	1
1	4	1
0	5	1
4	0	2
3	1	2
2	2	2
1	3	2
0	4	2
3	0	3
2	1	3
1	2	3
0	3	3
2	0	4
1	1	4
0	2	4
1	0	5
0	1	5
0	0	6

6

Random Composition of n into k Parts (RANCOM)

Our algorithm for random compositions is based on the "balls-in-cells" model which was described in the previous chapter. Briefly, we choose the positions of the cell boundaries at random, then by differencing we find out how many balls are in each cell.

The algorithm is quite fast, requiring just $O(k)$ operations per composition, on the average.

ALGORITHM RANCOM

(A) Choose $a_1, \ldots, a_{k-1}$, a random $(k-1)$-subset of $\{1, 2, \ldots, n+k-1\}$.

(B) Set $r_1 \leftarrow a_1 - 1$; $r_j \leftarrow a_j - a_{j-1} - 1$ $(j = 2, k-1)$; $r_k \leftarrow n + k - 1 - a_{k-1}$; Exit ∎

SUBROUTINE SPECIFICATIONS

(1) *Name of subroutine:* RANCOM.

(2) *Calling statement:* CALL RANCOM(N,K,R).

(3) *Purpose of subroutine:* Random composition of n into k parts.

(4) *Description of variables in calling statement:*

Name	Type	I/O/W/B	Description
N	INTEGER	*I*	Number whose compositions are desired.
K	INTEGER	*I*	Number of parts in desired composition.
R	INTEGER(K)	*O*	R(I) is the Ith part in the output composition (I=1,K).

(5) *Other routines which are called by this one:* RANKSB (Chapter 4), FUNCTION RAND(I) (random numbers).

(6) *Number of* FORTRAN *instructions:* 11.

```
      SUBROUTINE RANCOM(N,K,R)
      INTEGER R(K)
      CALL RANKSB(N+K-1,K-1,R)
      R(K)=N+K
      L=0
      DO 10 I=1,K
      M=R(I)
      R(I)=M-L-1
 10   L=M
      RETURN
      END
```

7

Next Permutation of n Letters (NEXPER)

There are many methods of sequentially producing all $n!$ permutations of n letters. Some of these methods require the construction of the next permutation ab initio while others produce it by a small modification of the previous permutation. Our inclination here is toward the latter approach because of its computational simplicity and elegance.

One knows, for example, that the transpositions generate the full permutation group, i.e., that each permutation σ of n letters is a product

$$\sigma = t_1 t_2 \cdots t_m$$

of transpositions. The question of algorithmic importance is this: can the $n!$ permutations be arranged in order in such a way that each one is obtainable from its predecessor by a single transposition?

For example, when $n = 3$ we have the list

$$
\begin{array}{ccc}
1 & 2 & 3 \\
1 & 3 & 2 \\
3 & 1 & 2 \\
3 & 2 & 1 \\
2 & 3 & 1 \\
2 & 1 & 3 \\
\end{array}
$$

in which each of the six permutations of three letters is obtained from its predecessor by a single exchange of two letters (transposition).

The same question can be asked in terms of graphs. Consider a graph of $n!$ vertices, one corresponding to each permutation of n letters. Let T_n denote the set of all transpositions. We construct a directed edge, in our graph, from vertex σ_1 to vertex σ_2 if there is a transposition $t \in T_n$ such that

$$\sigma_2 = t\sigma_1$$

For instance, in the case $n = 3$, the graph G has 6 vertices and looks like that shown in Fig. 7.1.

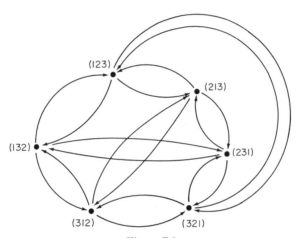

Figure 7.1

In terms of this graph, our question is just this: Is there a Hamilton path in the graph G? (A Hamilton path is a walk on the edges of G, following the "one-way" signs, which visits each vertex exactly once.) The question can also be asked about an arbitrary finite group G and set T of generators.

Problem 1 Given a finite group G and set T of generators of G. When can we conclude that all of the elements of G can be arranged in a sequence so that each one is obtainable from its immediate predecessor by the application of a single generator?

Problem 2 For which groups G can this be done for *every* set of generators of G? In Exercise 6 we see that S_5 is not such a group.

In the case at hand, the answer is always affirmative, i.e., the permutations of n letters can be arranged so that each is obtained by a single transposition from its predecessor on the list. In fact, several methods for doing this are known. An algorithm due to Wells [W1] accomplishes this in a nice way in that the amount of computational labor needed to decide which pair of letters to transpose at each stage is quite small. We mention also an elegant method of Trotter [Tr1] which sequences the transpositions so that at each step the two letters which are to be transposed are *adjacent* to each other. Trotter's method can be implemented by the following algorithm, which clearly reveals its inductive nature:

(A)   *[First entry]* $a_j \leftarrow j(j = 1, n)$; $m \leftarrow 1$; Exit.

(B)   *[Later entries]* $n' \leftarrow n$; $m' \leftarrow m$; $s \leftarrow n$.

(C)   *[Find n', the active letter]* $q \leftarrow m'$ (mod n'); $t \leftarrow m'$ (mod $2n'$); If $q \neq 0$, to **(D)**; If $t = 0$, set $s \leftarrow s - 1$; $m' \leftarrow m'/n'$; $n' \leftarrow n' - 1$; To **(C)**.

(D)   *[n' at left or right?]* If $q = t$, to **(E)**; $s \leftarrow s + q - n'$; To **(F)**.

(E)   *[Right end moves]* $s \leftarrow s - q$.

(F)   Exchange a_s, a_{s+1}; $m \leftarrow m + 1$; If $m = n!$, final exit; Exit ∎

Since the algorithms mentioned above are available in standard references, we present here another algorithm which has the advantage that it not only will produce a complete list of permutations if started at the identity permutation, but it can also be entered with any permutation and its sign, and will produce the successor of that permutation.

We describe the algorithm first recursively. Let $\mathscr{L}_n$ be the already constructed list of permutations on the letters $1, \ldots, n$, and let $\mathscr{L}_n\langle i \rangle$ $(1 \leq i \leq n + 1)$ be a list of permutations on the letters $1, \ldots, i - 1, i + 1, \ldots, n + 1$, defined as follows. $\mathscr{L}_n\langle n + 1 \rangle = \mathscr{L}_n$, and $\mathscr{L}_n\langle i \rangle$ $(1 \leq i \leq n)$ is obtained from $\mathscr{L}_n\langle i + 1 \rangle$ by replacing all occurrences of i by $i + 1$. Then

(1) $\mathscr{L}_{n+1} = \mathscr{L}_n\langle n + 1 \rangle \oplus (n + 1), \overline{\mathscr{L}_n\langle n \rangle} \oplus n, \mathscr{L}_n\langle n - 1 \rangle$

       $\oplus (n - 1), \overline{\mathscr{L}_n\langle n - 2 \rangle} \oplus (n - 2), \ldots$

The symbol $\oplus$ denotes that in $\mathscr{L}_n\langle i \rangle \oplus i$ each permutation has been extended to contain the letter i as last element. A bar, as in $\overline{\mathscr{L}\langle i \rangle}$, means reversal of the list. It is immediately clear, inductively, that this provides a listing in which each permutation differs from its predecessor by just one transposition; it is also clear, inductively, that each permutation occurs exactly once.

In order to transform this inductive definition into a noninductive one, and one in which we can compute the successor of a permutation without knowing its rank in the list (we only require its signature, to save the trouble of having to examine the whole permutation each time) we encode a permutation $(a_1, \ldots, a_n)$ by an $(n-1)$-tuple $(d_1, \ldots, d_{n-1})$, where d_i is the number of letters preceding a_{i+1} which are larger than a_{i+1}. This is essentially the well-known "inversion-table" of the permutation. Columns 4, 5, 6 in Table 7.1 give this encoding. Clearly, $0 \le d_i \le i$. It is easy to reconstruct a permutation from its encoding, starting with a_n and d_{n-1} and working forward. As most of the "action" is in the front, however, we want to calculate the successor of a permutation starting from the front. If only we know the signature, this can be done. If a permutation is even, we need no en- and decoding; we simply interchange a_1 and a_2. If the permutation is odd, however, we must operate more carefully. We consider the numbers $s_i = d_1 + \cdots + d_i$, and observe from

Table 7.1 The Permutations of $\{1, \ldots, 4\}$ and Their Inversion Tables

a_1	a_2	a_3	a_4	d_1	d_2	d_3
1	2	3	4	0	0	0
2	1	3	4	1	0	0
3	1	2	4	1	1	0
1	3	2	4	0	1	0
2	3	1	4	0	2	0
3	2	1	4	1	2	0
4	2	1	3	1	2	1
2	4	1	3	0	2	1
1	4	2	3	0	1	1
4	1	2	3	1	1	1
2	1	4	3	1	0	1
1	2	4	3	0	0	1
1	3	4	2	0	0	2
3	1	4	2	1	0	2
4	1	3	2	1	1	2
1	4	3	2	0	1	2
3	4	1	2	0	2	2
4	3	1	2	1	2	2
4	3	2	1	1	2	3
3	4	2	1	0	2	3
2	4	3	1	0	1	3
4	2	3	1	1	1	3
3	2	4	1	1	0	3
2	3	4	1	0	0	3

Table 7.1 that (starting from an *odd* permutation) if s_2 is odd, we find the code word for the successor by increasing d_2 by one unit, and if s_2 is even, we reduce d_2 by one unit. Exceptions to this rule occur when the operation on d_2 would push it outside the range $0 \le d_2 \le 2$. When that happens (e.g., at 3214) we consider s_3: if it is odd, we increase d_3 by one unit, otherwise we similarly reduce d_3, etc. The rigorous inductive proof of this statement follows by proving from (1) that the first element of $\mathscr{L}_n$ is encoded by a string of zeros, and the last element by all zeros except $d_{n-1} = n - 1$ if n is even, or $d_{n-2} = n - 2$, $d_{n-1} = n - 1$ if n is odd.

The in- or decreasing of d_i means, in practice, that we search among $a_1, \ldots , a_i$ for the nearest element (smaller or larger) to a_{i+1} in value, and interchange the two. Such an operation does, indeed, change *only* d_i.

ALGORITHM NEXPER

[Calculates the next permutation on n letters; successive permutations differ by only a transposition]

(A) *[First entry]* $a_i \leftarrow i$ $(i = 1, n)$; $\sigma \leftarrow 1$.
(B) *[Following entries; input $(a_1, \ldots , a_n)$ and signature σ]* If $\sigma = -1$, to **(C)**; $\sigma \leftarrow -1$; interchange a_1, a_2; Exit.
(C) $\sigma \leftarrow 1$; calculate $d_i = |\{j{:}j \le i,\ a_j > a_{i+1}\}|$ and $s_i = d_1 + \cdots + d_i$ $(i = 1, n - 1)$ until either s_i is odd and $d_i < i$ or s_i is even and $d_i > 0$; in the first (second) case search $a_1, \ldots , a_i$ for the largest (smallest) number less (greater) than a_{i+1}, and interchange the two; Exit ■

To estimate the amount of labor involved, we note that, to find the value of i in step **(C)**, $i(i + 1)/2$ steps are required, and finding the number with which to interchange a_i costs another i steps. In half the cases, none of this is needed, in the remaining $n!/2$ cases, there are $[(n!/2!) - (n!/3!)]$ for which $i = 2$, $[(n!/3!) - (n!/4!)]$ for which $i = 3$, etc. Hence, the total labor is

$$\sum_{i=2}^{n} \left(\frac{n!}{i!} - \frac{n!}{(i+1)!} \right) \frac{i(i+3)}{2} < n!(2e - 2)$$

Thus, the average labor per permutation is bounded, independent of n.

It is easy to see, inductively from (1), that the *last* permutation in the list $\mathscr{L}_n$ is

$$\begin{pmatrix} 1 & 2 & 3 & \cdots & n-1 & n \\ 2 & 3 & 4 & \cdots & n & 1 \end{pmatrix}$$

if n is even and

$$\begin{pmatrix} 1 & 2 & 3 & \cdots & n-2 & n-1 & n \\ 3 & 4 & 5 & \cdots & n & 2 & 1 \end{pmatrix}$$

if n is odd. The test for a final exit is done by checking the output permutation against the appropriate one of these.

SUBROUTINE SPECIFICATIONS

(1) *Name of subroutine:* NEXPER.
(2) *Calling statement:* CALL NEXPER (N,A,MTC,EVEN).
(3) *Purpose of subroutine:* Generate next permutation of 1, 2, . . . , n.
(4) *Descriptions of variables in calling statement:*

Name	Type	I/O/W/B	Description
N	INTEGER	I	Number of letters being permuted.
A	INTEGER(N)	I/O	A(I) is the value of the output permutation at I(I=1,N) (see Remarks below).
MTC	LOGICAL	I/O	=.TRUE. if current output is not the last permutation; =.FALSE. if no more permutations of n letters exist.
EVEN	LOGICAL	I/O	=.TRUE. if output permutation is even, =.FALSE. if it is odd. Not needed as input on first call, i.e., when MTC=.FALSE.

(5) *Other routines which are called by this one:* None.
(6) *Number of* FORTRAN *instructions:* 47.
(7) *Remarks:* If entered with MTC=.TRUE., any permutation A, and EVEN, the successor will be produced, unless A was the last permutation on N letters, in which case A(1) will be set to 0 on output.

```
SUBROUTINE NEXPER(N,A,MTC,EVEN)
INTEGER A(N),S,D
LOGICAL MTC,EVEN
IF(MTC) GO TO 10
```

```
        NM3=N-3
        DO 1  I=1,N
1       A(I)=I
        MTC=.TRUE.
5       EVEN=.TRUE.
        IF(N.EQ.1) GO TO 8
6       IF(A(N).NE.1.OR.A(1).NE.2+MOD(N,2)) RETURN
        IF(N.LE.3) GO TO 8
        DO 7  I=1,NM3
        IF(A(I+1).NE.A(I)+1) RETURN
7       CONTINUE
8       MTC=.FALSE.
        RETURN
10      IF(N.EQ.1) GO TO 27
        IF(.NOT.EVEN) GO TO 20
        IA=A(1)
        A(1)=A(2)
        A(2)=IA
        EVEN=.FALSE.
        GO TO 6
20      S=0
        DO 26  I1=2,N
25      IA=A(I1)
        I=I1-1
        D=0
        DO 30  J=1,I
30      IF(A(J).GT.IA) D=D+1
        S=D+S
        IF(D.NE.I*MOD(S,2)) GO TO 35
26      CONTINUE
27      A(1)=0
        GO TO 8
35      M=MOD(S+1,2)*(N+1)
        DO 40  J=1,I
        IF(ISIGN(1,A(J)-IA).EQ.ISIGN(1,A(J)-M)) GO TO 40
        M=A(J)
        L=J
40      CONTINUE
        A(L)=IA
        A(I1)=M
        EVEN=.TRUE.
        RETURN
        END
```

SAMPLE OUTPUT

The subprogram NEXPER was called repeatedly with N=5, until termination. The 120 output vectors A(1),A(2),...,A(5) follow.

```
1 2 3 4 5     2 5 1 3 4     1 5 4 3 2
2 1 3 4 5     5 2 1 3 4     5 1 4 3 2
3 1 2 4 5     3 2 1 5 4     4 1 5 3 2
1 3 2 4 5     2 3 1 5 4     1 4 5 3 2
2 3 1 4 5     1 3 2 5 4     1 3 5 4 2
3 2 1 4 5     3 1 2 5 4     3 1 5 4 2
4 2 1 3 5     2 1 3 5 4     5 1 3 4 2
2 4 1 3 5     1 2 3 5 4     1 5 3 4 2
1 4 2 3 5     1 2 4 5 3     3 5 1 4 2
4 1 2 3 5     2 1 4 5 3     5 3 1 4 2
2 1 4 3 5     4 1 2 5 3     4 3 1 5 2
1 2 4 3 5     1 4 2 5 3     3 4 1 5 2
1 3 4 2 5     2 4 1 5 3     1 4 3 5 2
3 1 4 2 5     4 2 1 5 3     4 1 3 5 2
4 1 3 2 5     5 2 1 4 3     3 1 4 5 2
1 4 3 2 5     2 5 1 4 3     1 3 4 5 2
3 4 1 2 5     1 5 2 4 3     2 3 4 5 1
4 3 1 2 5     5 1 2 4 3     3 2 4 5 1
4 3 2 1 5     2 1 5 4 3     4 2 3 5 1
3 4 2 1 5     1 2 5 4 3     2 4 3 5 1
2 4 3 1 5     1 4 5 2 3     3 4 2 5 1
4 2 3 1 5     4 1 5 2 3     4 3 2 5 1
3 2 4 1 5     5 1 4 2 3     5 3 2 4 1
2 3 4 1 5     1 5 4 2 3     3 5 2 4 1
2 3 5 1 4     4 5 1 2 3     2 5 3 4 1
3 2 5 1 4     5 4 1 2 3     5 2 3 4 1
5 2 3 1 4     5 4 2 1 3     3 2 5 4 1
2 5 3 1 4     4 5 2 1 3     2 3 5 4 1
3 5 2 1 4     2 5 4 1 3     2 4 5 3 1
5 3 2 1 4     5 2 4 1 3     4 2 5 3 1
5 3 1 2 4     4 2 5 1 3     5 2 4 3 1
3 5 1 2 4     2 4 5 1 3     2 5 4 3 1
1 5 3 2 4     3 4 5 1 2     4 5 2 3 1
5 1 3 2 4     4 3 5 1 2     5 4 2 3 1
3 1 5 2 4     5 3 4 1 2     5 4 3 2 1
1 3 5 2 4     3 5 4 1 2     4 5 3 2 1
1 2 5 3 4     4 5 3 1 2     3 5 4 2 1
2 1 5 3 4     5 4 3 1 2     5 3 4 2 1
5 1 2 3 4     5 4 1 3 2     4 3 5 2 1
1 5 2 3 4     4 5 1 3 2     3 4 5 2 1
```

8

Random Permutation of *n* Letters
(RANPER)

We produce a random permutation by a sequence of random interchanges. First choose any one of the n letters $1, \ldots, n$ for a_1; then choose any one of the remaining letters for a_2, etc. The construction of a permutation thus involves n choices, with respective probabilities

$$\frac{1}{n}, \frac{1}{n-1}, \ldots, \frac{1}{2}, 1$$

and the probability of a given permutation being chosen is therefore $1/n!$

ALGORITHM RANPER

(A) $a_i \leftarrow i \ (i = 1, n)$.
(B) For $m = 1, n : \{l \leftarrow m + \lfloor \xi(n + 1 - m) \rfloor\}$; Exchange $a_l, a_m\}$.
     Exit ■

The FORTRAN program contains a LOGICAL parameter SETUP. If it is set .FALSE. the subprogram will not setup the array A with $1, \ldots, n$ but, instead, will operate on whatever data the user has supplied.

SUBROUTINE SPECIFICATIONS

(1) *Name of subroutine:* RANPER.

(2) *Calling statement:* CALL RANPER(N,A,SETUP).

(3) *Purpose of subroutine:* Generate random permutation of *n* letters.

(4) *Descriptions of variables in calling statement:*

Name	Type	*I/O/W/B*	Description
N	INTEGER	*I*	Number of letters to be permuted.
A	INTEGER(N)	*O*	A(I) is the value of the output permutation at I (I=1,N).
SETUP	LOGICAL	*I*	If .FALSE., program shuffles the input array A; else, shuffles 1, 2, . . . , *n*.

(5) *Other routines which are called by this one:* Random number generator FUNCTION RAND(I).

(6) *Number of* FORTRAN *instructions:* 13.

```
      SUBROUTINE RANPER(N, A,SETUP)
      INTEGER A(N)
      LOGICAL SETUP
      IF(.NOT.SETUP) GO TO 20
      DO 10   I=1,N
10    A(I)=I
20    DO 40   M=1,N
30    L=M+RAND(1)*(N+1-M)
      L1=A(L)
      A(L)=A(M)
40    A(M)=L1
      RETURN
      END
```

SAMPLE OUTPUT

For each *n* = 3, . . . , 8, a set of 50 random permutations of *n* letters was chosen, and the number of cycles of each of these 50 permutations was found. In Table 8.1, we tabulate the average number

Table 8.1

n	(a)	(b)
3	1.78	1.83
4	2.04	2.08
5	2.28	2.28
6	2.50	2.45
7	2.52	2.59
8	2.58	2.72

of cycles in a permutation of n letters (a) estimated as described above and (b) calculated exactly from the formula $1 + \frac{1}{2} + \cdots + 1/n$.

9

Next Partition of Integer n (NEXPAR)

If n is a positive integer, then a representation

$$n = r_1 + r_2 + \cdots + r_k \quad (r_1 \geq r_2 \geq \cdots \geq r_k)$$

is called a *partition of n,* where it is understood that the "parts" $r_1, \ldots, r_k$ are strictly positive numbers. Thus, in Chapter 5, we saw that there are 28 *compositions* of 6 into 3 parts, but there are only 3 *partitions* of 6 into 3 parts, viz.,

$$
\begin{aligned}
6 &= 4 + 1 + 1 \\
&= 3 + 2 + 1 \\
&= 2 + 2 + 2
\end{aligned}
$$

Indeed, if we do not restrict the number of parts, then for a fixed n there are infinitely many *compositions* of n, but only a finite number of *partitions* of n. We let $p(n)$ denote the number of partitions of n. Then, for example, $p(6) = 11$ and the eleven partitions of 6 are

(1)
$$
\begin{aligned}
6 &= 6 \\
6 &= 5 + 1 \\
6 &= 4 + 2 \\
6 &= 4 + 1 + 1 \\
6 &= 3 + 3
\end{aligned}
$$

(cont.)

$$6 = 3 + 2 + 1$$
$$6 = 3 + 1 + 1 + 1$$
(1 cont.)
$$6 = 2 + 2 + 2$$
$$6 = 2 + 2 + 1 + 1$$
$$6 = 2 + 1 + 1 + 1 + 1$$
$$6 = 1 + 1 + 1 + 1 + 1 + 1$$

In the preceding list of partitions of 6, the arrangement of the partitions is in antilexicographic (reversed dictionary) order. More precisely, a partition

$$n = r_1 + r_2 + \cdots + r_k$$

occurs in the list above a partition

$$n = s_1 + s_2 + \cdots + s_q$$

if for some integer $t \geqq 0$ we have

(2) $\qquad r_i = s_i \quad (i = 1, \ldots, t) \qquad$ and $\qquad r_{t+1} > s_{t+1}$

The algorithm which we will now discuss generates from a given partition

(3) $\qquad\qquad n = r_1 + r_2 + \cdots + r_k$

its immediate successor on the list of all partitions of n, ordered antilexicographically. Suppose

(4) $\qquad\qquad n = \bar{r}_1 + \bar{r}_2 + \cdots + \bar{r}_l$

is the immediate successor. How can we determine the $\bar{r}_i$ from the r_i?

Suppose first that $r_k > 1$, e.g.,

(5) $\qquad\qquad 59 = 22 + 21 + 10 + 3 + 3$

What is the immediate successor of (5) in the list of partitions of 59? It is found by decreasing the last part r_k by 1 and adjoining a new part $= 1$:

(6) $\qquad\qquad 59 = 22 + 21 + 10 + 3 + 2 + 1$

Indeed, it is clear that (6) occurs *somewhere* after (5) on the list. If some third partition of 59 lies between (5) and (6), then it is easy to deduce a contradiction from the definition (2) of the ordering.

Hence, our first rule for obtaining (4) from (3) is:

(I) If $r_k > 1$,

set $\bar{r}_1 = r_1, \bar{r}_2 = r_2, \ldots, \bar{r}_{k-1} = r_{k-1}, \bar{r}_k = r_k - 1, \bar{r}_{k+1} = 1$;

Exit.

Now we need to deal with the case where $r_k = 1$. Suppose, in fact, that

$$r_k = r_{k-1} = \cdots = r_{j+1} = 1, \quad r_j > 1$$

as in the example

(7) $\qquad 59 = 19 + 16 + 14 + 3 + 1 + 1 + 1 + 1 + 1 + 1 + 1$

If the first part, 19, were to change to 18, say

(8) $\qquad\qquad\qquad 59 = 18 + \cdots$

we would not have the immediate successor of (7) because any partition

(9) $\qquad\qquad\qquad 59 = 19 + 15 + \cdots$

would lie between (7) and (8). Similarly, the 16 and the 14 must both remain fixed. Hence the immediate successor of (7) is of the form

(10) $\qquad\qquad 59 = 19 + 16 + 14 + \cdots$

and the dots in (10) constitute a partition of 10, namely, the one which is the successor of

$$10 = 3 + 1 + 1 + 1 + 1 + 1 + 1 + 1$$

in the list of partitions of 10. We need to go from the "last" partition of 10 whose largest part is 3 to the "first" partition of 10 whose largest part is 2

$$10 = 2 + 2 + 2 + 2 + 2$$

The successor of (7) is then

$$59 = 19 + 16 + 14 + 2 + 2 + 2 + 2 + 2$$

If we return now to the general case, suppose we have a partition

$$n = r_1 + r_2 + \cdots + r_j + 1 + 1 + 1 + \cdots + 1$$

In the successor partition, none of the first $j - 1$ parts will change, so that

$$\bar{r}_i = r_i \quad (i < j)$$

What remains is the last partition of the number

$$n' = r_j + (k - j)$$

whose largest part is r_j, and we must go to the first partition of n' whose largest part is

$$m = r_j - 1$$

This first partition is made by repeating the part m as often as it will "fit" into n', namely

$$\lfloor n'/m \rfloor$$

times, and if there is a positive remainder

$$s = n' - m\lfloor n'/m \rfloor$$

then we adjoin one additional part equal to s.

The complete formulation of the transition from (3) to (4) in the case where

$$r_{j+1} = \cdots = r_k = 1, \quad r_j > 1$$

is

$$
\begin{aligned}
\bar{r}_i &= r_i \quad (i = 1, \ldots, j - 1) \\
\bar{r}_j &= \bar{r}_{j+1} = \cdots = \bar{r}_{j+q-1} = m \\
\bar{r}_{j+q} &= s \quad (\text{if } s > 0)
\end{aligned}
$$

(II)

where $m = r_j - 1$, $\quad q = \lfloor (r_j + (k - j))/m \rfloor$, $\quad s = r_j + (k - j) - mq$.

The algorithm avoids the repeated listing of equal parts by maintaining a list $r_1 > r_2 > \cdots > r_d > 0$ of distinct parts, and a list $m_1, \ldots, m_d$ of their respective (positive) multiplicities. This economy results in a running time for each call to the subroutine which is independent of the value of n, and the program is loop-free.

ALGORITHM NEXPAR

(A) [*First entry*] $r_1 \leftarrow n$; $m_1 \leftarrow 1$; $d \leftarrow 1$; Exit.

(B) [*Later entries*] (Set σ equal to the sum of all parts of size one, plus the part preceding them.) If $r_d = 1$, set $\sigma \leftarrow m_d + 1$, $d \leftarrow d - 1$; Otherwise, $\sigma \leftarrow 1$.

(C) [*Remove one part of size r_d*] $f \leftarrow r_d - 1$; If $m_d = 1$, to (D); $m_d \leftarrow m_d - 1$; $d \leftarrow d + 1$.

(D) [*Add new parts of size f*] $r_d \leftarrow f$; $m_d \leftarrow \lfloor \sigma/f \rfloor + 1$.

(E) [*Add positive remainder*] $s \leftarrow \sigma \pmod{f}$; If $s = 0$, to (F); Otherwise, $d \leftarrow d + 1$; $r_d \leftarrow s$; $m_d \leftarrow 1$.

(F) [*Exit*] If $m_d = n$, final exit; Exit ∎

SUBROUTINE SPECIFICATIONS

(1) *Name of subroutine:* NEXPAR.
(2) *Calling statement:* CALL NEXPAR(N,R,M,D,MTC).
(3) *Purpose of subroutine:* Find next partition of *n*.
(4) *Descriptions of variables in calling statement:*

Name	Type	I/O/W/B	Description
N	INTEGER	I	Integer whose partitions are desired.
R	INTEGER(N)	I/O	R(I) is the Ith distinct part of the output partition (I=1,D).
M	INTEGER(N)	I/O	M(I) is the multiplicity of R(I) in the output partition (I=1,D).
D	INTEGER	I/O	Number of distinct parts in output partition.
MTC	LOGICAL	I/O	=.TRUE. if more partitions of N remain after this one. =.FALSE. if this is the last partition of N.

(5) *Other routines which are called by this one:* None.
(6) *Number of* FORTRAN *instructions:* 28.

```
      SUBROUTINE NEXPAR(N,R,M,D,MTC)
      IMPLICIT INTEGER (A-Z)
      LOGICAL MTC
      DIMENSION R(N),M(N)
      DATA NLAST/0/
10    IF(N.EQ.NLAST) GO TO 20
      NLAST=N
30    S=N
      D=0
50    D=D+1
      R(D)=S
      M(D)=1
40    MTC=M(D).NE.N
      RETURN
20    IF(.NOT.MTC) GO TO 30
      SUM=1
      IF(R(D).GT.1) GO TO 60
```

```
        SUM=M(D)+1
        D=D-1
  60    F=R(D)-1
        IF(M(D).EQ.1) GO TO 70
        M(D)=M(D)-1
        D=D+1
  70    R(D)=F
        M(D)=1+SUM/F
        S=MOD(SUM,F)
        IF(S) 40,40,50
        END
```

SAMPLE OUTPUT

The subprogram NEXPAR was called repeatedly with N=10, until termination. The 42 output partitions are shown below where, for clarity, we have shown multiple parts repeated.

```
10
 9  1
 8  2
 8  1  1
 7  3
 7  2  1
 7  1  1  1
 6  4
 6  3  1
 6  2  2
 6  2  1  1
 6  1  1  1  1
 5  5
 5  4  1
 5  3  2
 5  3  1  1
 5  2  2  1
 5  2  1  1  1
 5  1  1  1  1  1
 4  4  2
 4  4  1  1
 4  3  3
```

```
4  3  2  1
4  3  1  1  1
4  2  2  2
4  2  2  1  1
4  2  1  1  1  1
4  1  1  1  1  1  1
3  3  3  1
3  3  2  2
3  3  2  1  1
3  3  1  1  1  1
3  2  2  2  1
3  2  2  1  1  1
3  2  1  1  1  1  1
3  1  1  1  1  1  1  1
2  2  2  2  2
2  2  2  2  1  1
2  2  2  1  1  1  1
2  2  1  1  1  1  1  1
2  1  1  1  1  1  1  1  1
1  1  1  1  1  1  1  1  1  1
```

10

Random Partition of an Integer n (RANPAR)

For a given $n \geq 1$ we wish to select, uniformly at random (u.a.r.) a partition of n, i.e., so that each partition has a priori probability $1/p(n)$ of being selected.

Here is one method which suggests itself, and in fact, is an example of the ideas of Chapter 13 at work. Let $p(n, k)$ denote the number of partitions of n whose largest part is k. Then we first select the largest part of our partition according to

(1) $\qquad \text{Prob } \{r_1 = r\} = p(n, r)/p(n) \qquad (r = 1, \ldots, n)$

Next we need to choose u.a.r. a partition of n whose largest part is r_1.

Now observe that

(2) $\qquad p(n, k) = p(n - 1, k - 1) + p(n - k, k)$

because the first term on the right counts the partitions of n with largest part k and second part $< k$, while the second term counts those whose second part $= k$.

Thus to select a partition of n with largest part k:

(a) with probability $p(n - 1, k - 1)/p(n, k)$, select a partition of $n - 1$ whose largest part is $k - 1$ and add 1 to its largest part *or*

(b) with probability $p(n - k, k)/p(n, k)$, select a partition of $n - k$ whose largest part is k and make one more copy of its largest part.

See Chapter 13 for the full development of this method.

The above algorithm requires the tabulation of $p(n, k)$, and so next we describe an algorithm which avoids any tabulation of a function of two indices, requiring only a linear array.

Let $\sigma(n)$ denote the sum of the divisors of the integer n, and consider the identity

$$(3) \qquad np(n) = \sum_{m<n} \sigma(n - m)\, p(m)$$

of Euler. We give two proofs of this identity, the classical proof, and a purely combinatorial one which forms the basis of the present algorithm.

The original proof depends on the generating function

$$(4) \qquad \prod_{j=1}^{\infty} (1 - x^j)^{-1} = \sum_{n=0}^{\infty} p(n)x^n \quad (p(0) = 1)$$

whose proof can be found in any standard text. If we take logarithms of both sides and differentiate with respect to x, we obtain

$$(5) \qquad \sum_{j=1}^{\infty} \frac{jx^j}{1 - x^j} = \frac{\sum_n np(n)x^n}{\sum_n p(n)x^n}$$

If we develop the left side in a power series, we obtain

$$(6) \qquad \sum_{j=1}^{\infty} jx^j(1 + x^j + x^{2j} + \cdots) = \sum_{k=1}^{\infty} \sigma(k)x^k$$

and then (5) yields

$$(7) \qquad \left\{ \sum_{k=1}^{\infty} \sigma(k)x^k \right\} \left\{ \sum_{m=0}^{\infty} p(m)x^m \right\} = \sum_{n=0}^{\infty} np(n)x^n$$

Euler's identity now follows by equating the coefficients of x^n on both sides of (7).

Next we give a combinatorial proof of (3) which is based on a multiple counting of the partitions. Suppose π denotes a partition of some integer $m < n$ and d is a divisor of $n - m$; then the pair (π, d) gives rise to a partition of n. We simply adjoin to the partition π of m

exactly $(n - m)/d$ copies of d, yielding a partition π' of n. We make d copies of the partition π'.

We claim that as d runs over all divisors of $n - m$ and π runs over partitions of m ($m = 0, 1, \ldots, n - 1$), each partition of n is counted exactly n times by this process, which will prove (3). Indeed, if

$$(8) \qquad \pi' : n = \mu_1 r_1 + \cdots + \mu_k r_k$$

is a fixed partition of n where the r_i are the *distinct* parts of π' and the μ_i are their multiplicities, then for each t, $1 \leq t \leq \mu_i$, $1 \leq i \leq k$, π' is constructed by adjoining t copies of r_i to a partition of $n - tr_i$ and by replicating the resulting partition of n r_i times. This gives a total of

$$(9) \qquad \sum_{i=1}^{k} r_i \mu_i = n$$

copies of π' altogether, as required.

To obtain an algorithm for random partitions, we replace $n - m$ in (3) by jd, as we now describe. For simplicity of notation, we assume that $p(k) = 0$ for $k < 0$. Then

$$np(n) = \sum_{m=1}^{\infty} \sigma(m)p(n - m)$$

$$= \sum_{m=1}^{\infty} \sum_{d \mid m} dp(n - m)$$

$$= \sum_{j=1}^{\infty} \sum_{d=1}^{\infty} dp(n - jd)$$

or equivalently,

$$(10) \qquad 1 = \sum_{d=1}^{\infty} \sum_{j=1}^{\infty} \frac{dp(n - jd)}{np(n)}$$

and we interpret the terms on the right as probabilities which sum to 1.

ALGORITHM RANPAR

Given n.

(A) Set $\mathscr{P} \leftarrow$ empty partition; $m \leftarrow n$.

(B) Choose a pair of integers (d, j) according to the probabilities
$$\text{Prob}(d, j) = \frac{d p(m - jd)}{m p(m)} \quad (d, j = 1, 2, \ldots).$$

(C) Adjoin j copies of d to $\mathscr{P}$.

(D) $m \leftarrow m - jd$.

(E) If $m = 0$, exit; Otherwise go to **(B)** ∎

It is easy to see that all partitions of n have equal a priori probabilities of being chosen. Indeed, let

(11) $$n = \mu_1 d_1 + \mu_2 d_2 + \cdots + \mu_k d_k$$

be a fixed partition of n, where the $d_1, \ldots, d_k$ are distinct, and $\mu_1, \mu_2, \ldots$ are their multiplicities. This partition is chosen in $\mu_1 + \cdots + \mu_k$ ways by adjoining j copies of d_i to the partition of $n - jd_i$ given by

$$n - jd_i = \mu_1 d_1 + \cdots + (\mu_i - j) d_i + \cdots + \mu_k d_k$$
$$(1 \leqq j \leqq \mu_i; \ 1 \leqq i \leqq k)$$

Inductively, these latter partitions each have an a priori probability equal to $1/p(n - jd_i)$. The a priori probability of (11) is therefore

$$\sum_{i=1}^{k} \sum_{j=1}^{\mu_i} \frac{d_i p(n - d_i j)}{np(n)} \frac{1}{p(n - jd_i)} = \frac{1}{np(n)} \sum_{i=1}^{k} \sum_{j=1}^{\mu_i} d_i$$

$$= \frac{1}{np(n)} \sum_{i=1}^{k} \mu_i d_i$$

$$= \frac{1}{p(n)}$$

as required.

SUBROUTINE SPECIFICATIONS

(1) _Name of subroutine:_ RANPAR

(2) _Calling statement:_ CALL RANPAR(N,K,MULT,P).

(3) _Purpose of subroutine:_ Generate a random partition of n.

(4) _Descriptions of variables in calling statement:_

Name	Type	I/O/W/B	Description
N	INTEGER	I	Number whose partitions are desired.
K	INTEGER	O	Number of parts in output partition.
MULT	INTEGER(N)	O	MULT(I) is the multiplicity of I in the output partition (I=1,2,...,N).
P	INTEGER(N)	B	P(I) is the number of partition of I (I=1,N).

(5) *Other routines which are called by this one:* FUNCTION RAND(I) (random numbers).

(6) *Number of* FORTRAN *instructions:* 42.

```
        SUBROUTINE RANPAR(N,K,MULT,P)
        INTEGER P(N),D,MULT(N)
        DATA NLAST/0/
10      IF(N.LE.NLAST) GO TO 30
20      P(1)=1
        M=NLAST+1
        NLAST=N
        IF(N.EQ.1) GO TO 30
        DO 21  I=M,N
        ISUM=0
26      DO 22  D=1,I
        IS=0
        I1=I
24      I1=I1-D
        IF(I1) 22,25,23
23      IS=IS+P(I1)
        GO TO 24
25      IS=IS+1
22      ISUM=ISUM+IS*D
21      P(I)=ISUM/I
30      M=N
        K=0
        DO 31  I=1,N
31      MULT(I)=0
40      Z=RAND(1)*M*P(M)
        D=0
110     D=D+1
60      I1=M
        J=0
150     J=J+1
70      I1=I1-D
80      IF(I1)  110,90,120
120     Z=Z-D*P(I1)
130     IF(Z) 145,145,150
90      Z=Z-D
100     IF(Z) 145,145,110
145     MULT(D)=MULT(D)+J
        K=K+J
```

```
160   M=I1
170   IF(M.NE.0) GO TO 40
      RETURN
      END
```

SAMPLE OUTPUT

The subprogram RANPAR was called 880 times with N=6. The frequencies with which each of the 11 partitions of 6 were obtained are shown below. Thus, $6 = 3 + 2 + 1$ occurred 83 times, etc. The value $\chi^2 = 13.475$ was calculated from

$$\chi^2 = \sum_\pi \frac{(\phi(\pi) - 80)^2}{80}$$

where $\phi(\pi)$ is the frequency of the partition π, and the sum is over the 11 partitions of 6. In 95% of such experiments, the observed value of χ^2 would lie between 3.247 and 20.483 if the partitions did indeed have equal a priori probabilities.

```
 80   6
 77   5 1
106   4 2
 73   4 1 1
 72   3 3
 83   3 2 1
 67   3 1 1 1
 75   2 2 2
 76   2 2 1 1
 86   2 1 1 1 1
 85   1 1 1 1 1 1
```

CHI SQ IS 13.475 WITH 10 DEG FREEDOM

Postscript: Deus ex Machina

There is, in the material of Chapter 10, an excellent example of how new pure mathematics can result from the use of computers. In this case, we asked a question about the proper way to generate partitions at random. The mere asking of such a question was itself a product of the existence of computers. The resulting answer was obtained by seeking the combinatorial meaning of a certain identity (cf. Eq. (6)) and following the construction thereby suggested.

The next question is, "Why did it work?" That is, does our construction apply only to this problem, or is it a manifestation of something more general? It turns out that the latter is the case, and many other applications of the same ideas can be made (see Chapters 12 and 29).

The situation as regards partitions of an integer is that every partition is uniquely constructed from a set of basic building blocks, namely, the special partitions $1 = 1$, $2 = 2$, $3 = 3$, $4 = 4$, . . . (which we abbreviate as (1), (2), (3), (4), . . .) and their multiplicities. Thus

$$8 = 4 + 2 + 2 = (4) + 2 \cdot (2)$$

can be regarded as exhibiting a partition of 8 in terms of the special partitions (4), (2).

We have then a family of combinatorial objects, namely, the set of all partitions of all integers, and in that family is a distinguished subset (1), (2), (3), . . . , called "primes," with the property that every partition in the large family is uniquely expressible as a synthesis of primes with multiplicities. We shall now abstract from this case to a more general setting, and we will see that the algorithms follow along.

Consider a system $\mathcal{S}$ which consists of

1. a set $\mathcal{T}$ of objects
2. a *synthesis* map $\otimes : \mathcal{T} \times \mathcal{T} \to \mathcal{T}$
3. an *order function* $\Omega : \mathcal{T} \to Z^+$
4. a distinguished subset $\mathcal{P}$ of $\mathcal{T}$ of *primes*,

with the properties:

H1 Additivity of order under synthesis:

$$\Omega(t' \otimes t'') = \Omega(t') + \Omega(t'')$$

H2 Properties of synthesis: $\otimes$ is associative and commutative

H3 Unique factorization: There are no primes of order 0 and every $t \in \mathcal{T}$ is uniquely a synthesis of primes, i.e.,

$$t = p_1^{\mu_1} \otimes p_2^{\mu_2} \otimes \cdots$$

where $\forall_i : p_i \in \mathcal{P}$ and p^μ means $p \otimes p \otimes \cdots \otimes p$ (μ factors).

Let a_n denote the number of objects in $\mathcal{T}$ of order n, and let Π_n denote the number of *prime* objects of order n, for each $n = 1, 2, \ldots$. We seek the relationship between $\{a_n\}$ and $\{\Pi_n\}$.

Let $s \in \mathcal{T}$ be an object of order n, and let

(1) $$s = p_1^{\mu_1} \otimes p_2^{\mu_2} \otimes \cdots \otimes p_l^{\mu_l}$$

be the unique "prime factorization" of s. We shall construct the object s in all possible ways by a two-step process of synthesis and replication:

Step 1 (Synthesis) Let s' be any object of $\mathcal{T}$ whose prime decomposition is identical with Eq. (1) except that exactly *one* of the primes, say p_k, appears to a lower power, say $\mu_k - j$ ($1 \leq j \leq \mu_k$). Then

(2) $$s = s' \otimes p_k^j$$

The object s is uniquely determined by the object s' of lower order, and the integers j, p_k. Note that $\Omega(s') = \Omega(s) - j\Omega(p_k)$.

Step 2 (Replication) Make $d_k = \Omega(p_k)$ copies of the object s, after synthesizing it in Eq. (2).

Now, as s' runs over all objects of order $< n = \Omega(s)$, exactly how many copies of the object s will be made? For each j such that $1 \leq j \leq \mu_k$ we make $\Omega(p_k)$ copies of s, for a total of

$$\sum_{k=1}^{l} \mu_k \Omega(p_k) = \Omega(p_1^{\mu_1} \otimes p_2^{\mu_2} \otimes \cdots \otimes p_l^{\mu_1}) = \Omega(s) = n$$

copies of s altogether. Thus, *every object of order n is produced exactly n times*. It follows that

$$(3) \qquad n a_n = \sum_{j \geq 1} \sum_{d \geq 1} a_{n-jd} d \Pi_d$$

because the right side is the total number of copies of all objects of order n which are made by our synthesis and replication. The fundamental relation (3) expresses the total number of objects of each order in our system $\mathscr{S}$ in terms of the number of prime objects of each order.

The identity (3) can also be written in the more familiar equivalent form

$$(4) \qquad n a_n = \sum_{m < n} a_m \left\{ \sum_{d | (n-m)} d \Pi_d \right\}$$

The identity (6) of Chapter 10 appears as the analytic statement of unique factorization among the partitions of integers.

We can express (4) in terms of generating functions. Let

$$A(x) = \sum_{n \geq 0} a_n x^n, \qquad P(x) = \sum_{n \geq 1} \Pi_n x^n$$

be the counting functions for all objects and for prime objects, respectively, then

$$(5) \qquad A(x) = \exp \left\{ \sum_{r=1}^{\infty} \frac{P(x^r)}{r} \right\}$$

Indeed, logarithmic differentiation of (5), followed by matching coefficients of like powers of x yields (4) and conversely. (*Remark:* $a_0 = 1$; it counts the identity element, whose prime decomposition is the "empty" product.)

In this general situation, an algorithm for selecting an object of order n uniformly at random can always be given. Suppose that we know how to select a prime object uniformly of given order; then

(A) Choose a pair of integers (j, d) with probabilities

$$\text{Prob}(j, d) = \frac{d a_{n-jd} \Pi_d}{n a_n} \quad (j \geq 1, d \geq 1)$$

(B) Choose an object $s' \in \mathcal{T}$ of order $n - jd$ and a prime object p of order d uniformly.

(C) Synthesize s of order n from s' and j copies of p:

$$s = s' \otimes p^j$$

It is easy to check that every $s \in \mathcal{T}$, $\Omega(s) = n$, has a priori probability a_n^{-1} of being selected.

Aside from partitions of integers, there are many examples of such systems in the literature. If $\mathcal{T}$ is the set of all forests of rooted unlabeled trees, with $\Omega(t) = $ number of vertices of t, and if $\mathcal{P}$ is the subset of connected objects (i.e., rooted unlabeled trees), then we have the following interesting situation: Our algorithm will select a random forest on n vertices, given a knowledge of how to select a random tree of $m \leq n$ vertices. However, there is a 1–1 correspondence between forests of n vertices and trees of $n + 1$ vertices: just adjoin a new vertex $n + 1$, call it the new root, and connect it to each of the roots of the trees in the original forest. It follows that our algorithm will select a *tree* of $n + 1$ vertices if we know how to select a *tree* of $\leq n$ vertices, i.e., we have an *inductive* process for selecting random rooted, unlabeled trees! This is discussed in detail in Chapter 29.

The abstract structures which we have introduced here are a special case of a family of objects, called "prefabs," first studied by E. Bender and J. Goldman who also pointed out the interesting property of rooted forests noted above. What we have added to their work is first of all the constructive procedure of synthesis and replication, which gives direct combinatorial meaning to the logarithmic derivative (4) of the functional equation (5), and secondly the realization that such structures are invariably equipped with recurrent algorithms for randomly uniform selection of objects of given order.

We give one more example of a prefab, from the theory of plane partitions. By a plane partition of n we mean an array $n_{ij}]_{i,j=1}^{\infty}$ of nonnegative integers such that

(6)

$$\text{(a)} \quad \sum_{i,j} n_{ij} = n$$

$$\text{(b)} \quad \begin{array}{l} n_{ij} \geq n_{i,j+1} \\ n_{ij} \geq n_{i+1,j} \end{array} \quad (i, j = 1, 2, \ldots)$$

Thus, for example, the array

(7)
$$
\begin{array}{cccccc}
7 & 3 & 2 & 2 & 2 & 1 \\
4 & 2 & 1 & 1 & & \\
2 & 2 & & & &
\end{array}
$$

is a plane partition of 29, in which the blank entries are all zero.

We will show that the set of all plane partitions is a prefab, in which the order of a plane partition of n is n, and in which we will describe the synthesis operation and identify the "prime" partitions. We will see that there are exactly n prime partitions of each order n, and of course each partition will be uniquely expressible as a "product" of these.

When this has been demonstrated, it will follow from Eq. (5) of this chapter, which holds in any prefab, that

$$
\begin{aligned}
A(x) &= \exp \left\{ \sum_{r=1}^{\infty} \frac{1}{r} \sum_{n=1}^{\infty} n x^{rn} \right\} \\
&= \exp \left\{ \sum_{n=1}^{\infty} n \sum_{r=1}^{\infty} \frac{x^{rn}}{r} \right\} \\
&= \exp \left\{ \sum_{n=1}^{\infty} n \log \frac{1}{(1 - x^n)} \right\} \\
&= \frac{1}{\displaystyle\prod_{n=1}^{\infty} (1 - x^n)^n}
\end{aligned}
$$

Thus, the generating function for the number, $b(n)$, of plane partitions of n, is

(8)
$$
\sum_{n \geq 0} b(n) x^n = \frac{1}{\displaystyle\prod_{n=1}^{\infty} (1 - x^n)^n}
$$

The difficult step, then, is to identify the synthesis operation and the prime objects. Now, the following result is due to Bender and Knuth [BK 1].

Theorem There is a one-to-one correspondence between *plane partitions of n*, on the one hand, and *infinite matrices $a_{ij}(i, j \geq 1)$ of nonnegative integer entries which satisfy*

(9)
$$
\sum_{r \geq 1} r \left\{ \sum_{i+j=r+1} a_{ij} \right\} = n
$$

on the other.

Let us define Γ_n to be the set of all infinite matrices of nonnegative integer entries that satisfy (9), and let

$$\Gamma = \bigcup_{n \geq 1} \Gamma_n$$

Then, for example, there are exactly six plane partitions of 3, namely,

$$
\begin{array}{ccccc}
3 & 21 & 2 & 111 & 11 & 1 \\
   &   & 1 &   & 1 & 1 \\
   &   &   &   &   & 1
\end{array}
$$

and there are exactly six matrices in Γ_3, namely,

$$
\begin{array}{cccccc}
3 & 10 & 11 & 00 & 001 & 000 \\
  & 10 & 00 & 01 & 000 & 000 \\
  &   &   &   & 000 & 100
\end{array}
$$

in which matrix elements that are not shown are all zeros.

In view of the theorem, we can consider plane partitions of n to be encoded forms of the matrices of Γ_n. Therefore, in order to generate all plane partitions of n or to select a random plane partition of n, we may generate all matrices of Γ_n, or select one at random, and then decode the resulting output so as to present the plane partition in familiar form.

Now we assert that the set Γ of matrices is a prefab. Indeed, if $A \in \Gamma$, the order $\Omega(A)$ of A is defined to be the left-hand side of (9). The synthesis operation $\otimes$ in Γ is ordinary matrix addition. We observe that

$$\Omega(A \otimes B) = \Omega(A + B) = \Omega(A) + \Omega(B)$$

What are the primes of Γ? They are the matrices A all of whose entries are 0 except for a single 1 in one position.

To be a prime of order n a matrix will have its single 1 entry on the nth antidiagonal. Hence there are exactly n prime objects of order n for each $n \geq 1$. Evidently each $A \in \Gamma$ is uniquely expressible as a synthesis (= sum) of primes.

Let us denote by σ the bijection between Γ and the plane partitions, whose existence is asserted by the theorem of Bender–Knuth, so that $\sigma(A)$ is the plane partition associated with the matrix $A \in \Gamma$.

It follows from the earlier considerations of this Postscript that we can select a plane partition of n uniformly at random by selecting a matrix $A \in \Gamma_n$ u.a.r. and decoding it by the mapping σ, which we describe below. The selection of A follows our previous prescription exactly, and aside from the decoding problem, the preparation of a computer program to do so would closely parallel our program RANPAR, and so is left as an exercise for the reader.

Another consequence of the theorem of Bender–Knuth is that we can generate sequentially all plane partitions of n. To do this, we generate the matrices A of Γ_n and decode them by σ. According to (9), which we must do is the following.

ALGORITHM NEXT PLANE PARTITION

For each linear partition of n, π: $n = \mu_1 + 2\mu_2 + 3\mu_3 + 4\mu_4 + \ldots$ do:
   For each set of compositions of μ_1 into 1 part, μ_2 into 2 parts, . . . , μ_j into j parts, . . . , do:
      Enter the parts of the composition of μ_r into the rth antidiagonal of the matrix A ($r = 1, 2, \ldots$)
   End
End ■

The problem is therefore an easy application of NEXPAR (Chapter 9) and NEXCOM (Chapter 5) and is also left to the reader.

It remains to describe the bijection σ. We proceed in four steps, beginning with a matrix $A \in \Gamma_n$ and ending with a plane partition $\sigma(A)$ of n.

First, from $A \in \Gamma_n$ we construct a two-line array $\sigma_1(A)$. Precisely, suppose $a_{ij} = m > 0$. Then enter m copies of i in the first row of $\sigma_1(A)$ and m copies of j in the second row. For example, if we start with

$$(10) \qquad A = \begin{pmatrix} 1 & 0 & 2 \\ 0 & 2 & 0 \\ 1 & 0 & 0 \end{pmatrix}$$

we would obtain

$$(11) \qquad \sigma_1(A) = \begin{pmatrix} 1 & 1 & 1 & 2 & 2 & 3 \\ 1 & 3 & 3 & 2 & 2 & 1 \end{pmatrix}$$

Second, we permute the columns of $\sigma_1(A)$ so that (a) the elements of the first row are in nonincreasing order and (b) within a block of constancy of the first row, the corresponding elements of the second row are in nonincreasing order. This yields $\sigma_2(A)$. In the preceding example we have

(12)
$$\sigma_2(A) = \begin{pmatrix} 3 & 2 & 2 & 1 & 1 & 1 \\ 1 & 2 & 2 & 3 & 3 & 1 \end{pmatrix}$$

Third, from the two-line array

$$\sigma_2(A) = \begin{pmatrix} i_1, & i_2, & \cdots, & i_m \\ j_1, & j_2, & \cdots, & j_m \end{pmatrix}$$

we construct a *pair* S, T of plane partitions by an insertion and bumping procedure, as follows. The plane partition S will be con-constructed from $i_1, \ldots, i_m$ and T from $j_1, \ldots, j_m$. Recursively, define $S^{(1)} = i_1$ and $T^{(1)} = j_1$. Suppose that $S^{(r)}$ and $T^{(r)}$ have been constructed, and that these are plane partitions of the same shape, $S^{(r)}$ containing the parts $i_1, \ldots, i_r$ and $T^{(r)}$ containing $j_1, \ldots, j_r$.

We then insert j_{r+1} into the first row of $T^{(r)}$, immediately to the right of the rightmost entry which is $\geq j_{r+1}$. If this space is occupied by some element k, then by entering j_{r+1} into this space we bump k down to the second row, where it is then treated just as j_{r+1} was, so that another element may be bumped to the third row, etc. If there is no entry that is $\geq j_{r+1}$, then j_{r+1} is inserted at the beginning of the row and bumps the former first element down.

In this way, $T^{(r+1)}$ is formed from $T^{(r)}$ and j_{r+1}. To construct $S^{(r+1)}$ is easy: just insert i_{r+1} into $S^{(r)}$ so that the resulting array has the same shape as $T^{(r+1)}$.

Let us follow the matrix A of our example through this process.

$T^{(r)}$	$S^{(r)}$
1	3
2	3
1	2
22	32
1	2
32	32
2	2
1	1
33	32
22	21
1	1
331	321
22	21
1	1

Thus the pair of plane partitions which correspond to A is

(13)
$$
\begin{matrix}
& 331 & & 321 \\
S = & 22 & T = & 21 \\
& 1 & & 1
\end{matrix}
$$

and this completes the third phase of the construction.

Finally, from the ordered pair S, T of plane partitions we construct a single plane partition $\sigma = \sigma(A)$ by a method of Frobenius, as adapted by Bender and Knuth.

From a column of S and a column of T we form a new column, as illustrated below. The labels at the right of the array are the first column of S, those at the bottom of the array are *one less* than the first column of T.

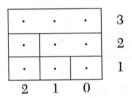

By counting the dots in each row we find the first column of the plane partition σ, namely,

$$
\begin{matrix}
3 \\
3 \\
3
\end{matrix}
$$

Repeat this with the second column of S and of T to obtain

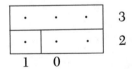

from which the second column of the plane partition is

$$
\begin{matrix}
3 \\
3
\end{matrix}
$$

Finally from the third columns of S and T we get

and the third column of σ is 1. Thus, the complete partition is

$$\sigma(A) = \sigma = \begin{array}{ccc} 3 & 3 & 1 \\ 3 & 3 & \\ 3 & & \end{array}$$

Note that this is a plane partition of 16 as required by (9), (10). The reader is encouraged to write the subroutines NXPLPR and RNPLPR which, respectively, sequence all plane partitions of a given integer n and select such a partition uniformly at random, as well as to write DCDPLP which decodes the matrix output of the previous two programs so as to output the familiar form of a plane partition.

11

Next Partition of an n-Set (NEXEQU)

From partitions of integers we turn to partitions of sets. If $S = \{1, 2, \ldots , n\}$, then by a *partition* of S we mean a family of sets $T_1, T_2, \ldots , T_k$ satisfying

(a) $$T_i \cap T_j = \varnothing \quad (i \neq j)$$

and

(b) $$\bigcup_{i=1}^{k} T_i = S$$

and

(c) $$T_i \neq \varnothing \quad (i = 1, \ldots , k)$$

It is assumed that no significance is attached to the order in which $T_1, \ldots , T_k$ are listed nor to the order of the listing of elements within these sets. For $n = 3$, therefore, we have the following 5 partitions:

$$
\begin{array}{ll}
1 & (123) \\
2 & (12)\ (3) \\
3 & (13)\ (2) \\
4 & (23)\ (1) \\
5 & (1)\ (2)\ (3)
\end{array}
$$

A partition of a set is evidently identical with an equivalence relation on the set, with the T_i as the equivalence classes, which accounts for the name NEXEQU of this routine.

Given a partition $\mathscr{P}$ of $\{1, 2, \ldots, n\}$ into k classes $T_1, \ldots, T_k$, we may associate with $\mathscr{P}$ exactly $k + 1$ different partitions of $\{1, 2, \ldots, n, n + 1\}$; namely,

$$\mathscr{P}_1 : T_1 \cup \{n + 1\}, T_2, T_3, \ldots, T_k$$
$$\mathscr{P}_2 : T_1, T_2 \cup \{n + 1\}, T_3, \ldots, T_k$$
$$\vdots$$
$$\mathscr{P}_k : T_1, T_2, T_3, \ldots, T_k \cup \{n + 1\}$$
$$\mathscr{P}_{k+1} : T_1, T_2, T_3, \ldots, T_k, \{n + 1\}$$

The first k of these descendants of $\mathscr{P}$ have k classes, and the last one has $k + 1$ classes in which the last class is just the singleton $\{n + 1\}$.

We can visualize all of the partitions of $\{1, 2, \ldots, n\}$, therefore, as one horizontal line in a tree (Fig. 11.1).

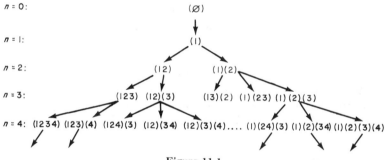

Figure 11.1

In this tree, a line is drawn from each partition $\mathscr{P}$, of a set of n letters into k classes, to each of its $k + 1$ immediate descendants, which are partitions of $n + 1$ letters into k or $k + 1$ classes.

To generate successively all partitions of $\{1, 2, \ldots, n\}$ we want to move from left to right along one horizontal level of this tree. Given a partition $\mathscr{P}$ of $\{1, 2, \ldots, n\}$, to locate its immediate successor $\mathscr{P}'$ we (a) locate the highest active letter m, i.e., the largest integer $1 \leqq m \leqq n$ which is not in a singleton class in $\mathscr{P}$; (b) move m to the next higher class, or create a singleton class for m if m is already in the highest class; (c) put $m + 1, \ldots, n$ into class 1.

For example, here are the five partitions of $\{1, 2, 3\}$, in the order in which this algorithm produces them where, in each case, the highest active letter m is underlined:

$$(1 \ 2 \ \underline{3})$$
$$(1 \ \underline{2}) \ (3)$$
$$(1 \ \underline{3}) \ (2)$$
$$(1) \ (2 \ \underline{3})$$
$$(1) \ (2) \ (3)$$

It is convenient to use two arrays, p_i, the population of the ith class, and q_j, the class to which element j belongs $(i = 1, \ n_c; \ j = 1, \ n)$.

ALGORITHM NEXEQU

(A)   *[First entry]* $p_1 \leftarrow n; \ n_c \leftarrow 1; \ q_i \leftarrow 1 \ (i = 1, \ n);$ to (**F**).

(B)   *[Later entries]* $m \leftarrow n.$

(C)   *[Find highest active letter m]* $l \leftarrow q_m;$ If $p_l \neq 1,$ to (**D**); $q_m \leftarrow 1; \ m \leftarrow m - 1;$ to (**C**).

(D)   *[Move m]* $n_c \leftarrow n_c + m - n; \ p_1 \leftarrow p_1 + n - m;$ If $l \neq n_c,$ to (**E**); $n_c \leftarrow n_c + 1; \ p(n_c) \leftarrow 0.$

(E) $q_m \leftarrow l + 1; \ p_l \leftarrow p_l - 1; \ p_{l+1} \leftarrow p_{l+1} + 1.$

(F)   If $n_c = n,$ final exit; Exit ■

Remark   Recall that no significance attaches to the names of the classes within a partition. Yet, on output, the computer is forced to name these classes for reference purposes. Our algorithm does this naming in such a way that the smallest element of class j is the first integer which does not belong to any earlier class $1, 2, \ldots , j - 1$.

SUBROUTINE SPECIFICATIONS

(1)   *Name of subroutine:*   NEXEQU.

(2)   *Calling statement:*   CALL NEXEQU(N,NC,P,Q,MTC).

(3)   *Purpose of subroutine:*   Generate next equivalence relation on $\{1, 2, \ldots , n\}$.

(4)   *Descriptions of variables in calling statement:*

Name	Type	I/O/W/B	Description
N	INTEGER	I	Number of elements in set to be partitioned.
NC	INTEGER	O	Number of classes in output partition.
P	INTEGER(N)	I/O	P(I) is the number of elements in the Ith class of the output partition (I=1,NC).
Q	INTEGER(N)	I/O	Q(I) is the class to which I belongs (I=1,N).
MTC	LOGICAL	I/O	=.TRUE. if current output is not the last; =.FALSE. otherwise.

(5) *Other routines which are called by this one:* None.
(6) *Number of* FORTRAN *instructions:* 26.

```
      SUBROUTINE NEXEQU(N,NC,P,Q,MTC)
      INTEGER P(N),Q(N)
      LOGICAL MTC
      IF(MTC) GO TO 20
10    NC=1
      DO 11   I=1,N
11    Q(I)=1
      P(1)=N
60    MTC=NC.NE.N
      RETURN
20    M=N
30    L=Q(M)
      IF(P(L).NE.1) GO TO 40
      Q(M)=1
      M=M-1
      GO TO 30
40    NC=NC+M-N
      P(1)=P(1)+N-M
      IF(L.NE.NC) GO TO 50
      NC=NC+1
      P(NC)=0
50    Q(M)=L+1
      P(L)=P(L)-1
      P(L+1)=P(L+1)+1
      GO TO 60
      END
```

SAMPLE OUTPUT

The subprogram NEXEQU was called repeatedly with N=5, until termination. The 52 output vectors $Q(1),Q(2),\ldots,Q(5)$ are shown below.

1	1	1	1	1		1	2	2	1	2
1	1	1	1	2		1	2	2	1	3
1	1	1	2	1		1	2	2	2	1
1	1	1	2	2		1	2	2	2	2
1	1	1	2	3		1	2	2	2	3
1	1	2	1	1		1	2	2	3	1

1	1	2	1	2		1	2	2	3	2
1	1	2	1	3		1	2	2	3	3
1	1	2	2	1		1	2	2	3	4
1	1	2	2	2		1	2	3	1	1
1	1	2	2	3		1	2	3	1	2
1	1	2	3	1		1	2	3	1	3
1	1	2	3	2		1	2	3	1	4
1	1	2	3	3		1	2	3	2	1
1	1	2	3	4		1	2	3	2	2
1	2	1	1	1		1	2	3	2	3
1	2	1	1	2		1	2	3	2	4
1	2	1	1	3		1	2	3	3	1
1	2	1	2	1		1	2	3	3	2
1	2	1	2	2		1	2	3	3	3
1	2	1	2	3		1	2	3	3	4
1	2	1	3	1		1	2	3	4	1
1	2	1	3	2		1	2	3	4	2
1	2	1	3	3		1	2	3	4	3
1	2	1	3	4		1	2	3	4	4
1	2	2	1	1		1	2	3	4	5

12

Random Partition of an n-Set (RANEQU)

The algorithm for a random partition of the set $\{1, 2, \ldots, n\}$ follows the basic idea of Chapter 10 and its Postscript, in which we found a recurrence formula and then endowed it with a probabilistic interpretation.

Here, the recurrence is in the quantities $a_0, a_1, \ldots$, where a_n is the number of partitions of a set of n objects. Indeed, let $\mathscr{P}_k$ be a fixed partition of $\{1, 2, \ldots, k\}$. We will extend $\mathscr{P}_k$ to exactly $\binom{n-1}{k}$ partitions of $\{1, 2, \ldots, n\}$. First, choose a subset S of k elements from $\{1, 2, \ldots, n-1\}$. Relabel the k elements of $\mathscr{P}_k$ using the elements of S as labels and preserving order. Adjoin to the resulting partition all of the $n - k$ remaining elements of $\{1, 2, \ldots, n\}$ regarded as a single class. In this way, we make $\binom{n-1}{k}$ partitions of $\{1, 2, \ldots, n\}$ from each of the a_k partitions of $\{1, 2, \ldots, k\}$ or

$$\sum_{k=0}^{n-1} \binom{n-1}{k} a_k$$

partitions of $\{1, 2, \ldots, n\}$ altogether.

We claim that every partition $\mathscr{P}$ of $\{1, 2, \ldots, n\}$ is constructed just once in this way. For, if $\mathscr{P} = T_1 \cup T_2 \cup \cdots \cup T_h$ where the T_i are the classes of $\mathscr{P}$, suppose T_h is the unique class of $\mathscr{P}$ which contains the element n. Then $\mathscr{P}$ was constructed uniquely from the partition which is formed by relabeling the set $T_1 \cup T_2 \cup \cdots \cup T_{h-1}$ with labels $1, 2, \ldots, k$, where $k = \Sigma_{i=1}^{h-1} |T_i|$. Hence we have

$$(1) \qquad a_n = \sum_{k=0}^{n-1} \binom{n-1}{k} a_k \quad (n \geq 1,\ a_0 = 1)$$

from which we find $a_1 = 1$, $a_2 = 2$, $a_3 = 5$, $a_4 = 15$, etc.

Usually the use to which (1) is put is to find the generating function

$$(2) \qquad f(z) = \sum_{n=0}^{\infty} \frac{a_n}{n!} z^n$$

In fact, (1) implies that

$$f'(z) = e^z f(z)$$

and $f(0) = 1$. The solution of this initial value problem is evidently

$$(3) \qquad f(z) = \exp(e^z - 1)$$

which identifies the numbers a_n as

$$(4) \qquad a_n = \frac{d^n}{dz^n} (\exp(e^z - 1)) \Big|_{z=0} \quad (n = 0, 1, 2, \ldots)$$

The numbers $a_1, a_2, \ldots$ are called the *Bell numbers*.

Our interest, however, lies in the algorithm which results from dividing both sides of (1) by a_n

$$(5) \qquad 1 = \sum_{k=0}^{n-1} \binom{n-1}{k} \frac{a_k}{a_n}$$

and identifying the terms on the right as probabilities which sum to 1.

Indeed, the kth term of the sum is the probability that the class in which n lives contains exactly k other elements besides n.

Now we could use a straightforward algorithm which starts by choosing the companions for n, relabeling the remaining elements, etc. However, a simpler method avoids all relabelings and reduces the bookkeeping to a minimum.

First we will choose the sizes $k_1, k_2, \ldots, k_l$ of the classes in the output partition by sampling them from the relevant distribution. Next we set up an array of length n. In the last k_1, places we insert 1's in this array; in the next k_2 places we put 2's, $\ldots$, in the first k_l places we insert l's. Finally we execute a random permutation of this array and exit, at which time the ith array element is interpreted as the class to which the letter i belongs. We give below a formal algorithm and then a proof of its validity. It will be noted that the initial insertion of class numbers into the array is done while the sizes are being chosen, so that the k_i themselves need never be stored.

ALGORITHM RANEQU

[Produces a random equivalence relation on $\{1, \ldots, n\}$. The output is $(q_1, \ldots, q_n)$, where q_i is the number of the class to which i belongs.]

(A) [*Initialization*] Precalculate and store any a_i not yet calculated (see(1)); $m \leftarrow n$; $l \leftarrow 0$.
(B) [*Choose sizes of equivalence classes*] Choose k according to the probability

(6) $$\text{Prob}(k) = \binom{m-1}{k-1} \frac{a_{m-k}}{a_m} \quad (1 \leq k \leq m)$$

$l \leftarrow l + 1$; store l into $q_{m-k+1}, \ldots, q_m$; $m \leftarrow m - k$; if $m > 0$, to (B).
(C) [*Randomize*] Perform a random permutation on $(q_1, \ldots, q_n)$; Exit ∎

To prove the validity of the algorithm we modify it slightly, in a way which gives the same result, and is more transparent, but requires an additional array of storage $(c_1, \ldots, c_n)$. We choose a random permutation π (the same as in step (C)), and store its inverse (which is equally probable) into $(c_1, \ldots, c_n)$. Then we choose k according to (6), with $m = n$, and store 1 into $q_{n-k+1}, \ldots, q_n$. Next, proceed with equivalence class 2, etc. Now the equivalence classes are the same as in the algorithm; that is: in the algorithm, q_i is the equivalence class of i, while in the above modification q_i is the equivalence class of c_i.

Now, let $\mathscr{P}$ be a fixed partition of $\{1, \ldots, n\}$, with classes $T_1, \ldots, T_h$. Then the probability that $\mathscr{P}$ is obtained by the modified algorithm is a sum of probabilities ΣP_i, where P_i is the probability that in the first step *one* of $T_1, \ldots, T_h$ is found, namely T_i, multiplied by the probability that the remaining partition $\mathscr{P} - T_i = \{T_j\}_{j \neq i}$ is then obtained as an equivalence relation on $\{1, \ldots, n\} - T_i$. By induction on n we may assume that the latter is $1/a_{n-k_i}$, where $k_i = |T_i|$. The probability that k_i is first found is $\text{Prob}(k_i)$ (see (6)), with $m = n$; the probability that $c_{n-k_i+1}, \ldots, c_n$ are the elements of T_i, in some order, is $\binom{n}{k_i}^{-1}$. Thus we find

$$\text{Prob}(\mathscr{P}) = \sum_{i=1}^{h} \text{Prob}\{k = |T_i|\} \cdot \text{Prob}\{\{c_{n-k+1}, \ldots, c_n\} = T_i\}$$

$$\cdot \text{Prob}\{\mathscr{P} - T_i\}$$

$$= \sum_{i=1}^{h} \binom{n-1}{k_i-1} \frac{a_{n-k_i}}{a_n} \binom{n}{k_i}^{-1} \frac{1}{a_{n-k_i}} = \sum_{i=1}^{h} \frac{k_i}{na_n} = \frac{1}{a_n} \; \blacksquare$$

The flow chart and the FORTRAN program are basically a straightforward implementations of the algorithm. However, a few remarks may be helpful. In order to control the order of magnitude of the quantities, we actually do not precompute the a_i, but, instead $b_i = a_i/i!$. In the evaluation of the sum in the program (this was deleted from the flow chart)

$$nb_n = \sum_{k=0}^{n-1} \frac{1}{k!} b_{n-k-1}$$

we use a nested multiplication to build up $k!$. Also, we do not test against a random number ξ between 0 and 1, but against $\xi m b_m$ (Box 20) thus saving a division in each step. In the testing, another nested factorial is hidden in three instructions in Box 40.

FLOW CHART RANEQU

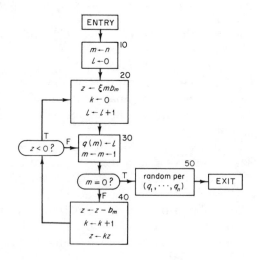

SUBROUTINE SPECIFICATIONS

(1) *Name of subroutine:* RANEQU.

(2) *Calling statement:* CALL RANEQU(N,L,Q,B).

(3) *Purpose of subroutine:* Generate random equivalence relation on $\{1, 2, \ldots, n\}$.

(4) *Descriptions of variables in calling statement:*

Name	Type	I/O/W/B	Description
N	INTEGER	I	Cardinality of partitioned set.
L	INTEGER	O	Number of classes in output partition.
Q	INTEGER	O	Q(I) is the class to which I belongs (I=1,N).
B	REAL(N)	B	B(K)=A(K)/K! (see text Eq. (1)) (K=1,N).

(5) *Other routines which are called by this one:* FUNCTION RAND(I) (random numbers), RANPER.

(6) *Number of FORTRAN instructions:* 30.

(7) *Remark:* Equivalence classes numbered in no particular order.

```
      SUBROUTINE RANEQU(N,L,Q,B)
      INTEGER Q(N)
      REAL B(N)
      DATA NLAST/1/
      B(1)=1.
      IF(N .LE. NLAST) GO TO 10
      NM1=N-1
      DO 5 L=NLAST,NM1
      SUM=1./L
      L1=L-1
      IF(L1 .EQ. 0) GO TO 5
      DO 6 K=1,L1
6     SUM=(SUM+B(K))/(L-K)
5     B(L+1)=(SUM+B(L))/(L+1)
      NLAST=N
10    M=N
      L=0
20    Z=M*B(M)*RAND(1)
      K=0
      L=L+1
30    Q(M)=L
      M=M-1
```

```
      IF (M .EQ. 0) GO TO 50
40    Z=Z-B(M)
      K=K+1
      Z=Z*K
      IF (Z) 20,30,30
50    CALL RANPER(N,Q,.FALSE.)
      RETURN
      END
```

SAMPLE OUTPUT

The subprogram RANEQU was called 750 times with N=4. The frequencies with which each of the 15 partitions of the set $\{1, 2, 3, 4\}$ occurred are shown below. The partitions are identified by the output vector $Q: Q(I)$ is the class to which I belongs $(I=1,4)$. Thus, the partition (124) (3) was obtained 46 times, etc. The value $\chi^2 = 6.64$ was calculated from

$$\chi^2 = \sum_{\pi} \frac{(\phi(\pi) - 50)^2}{50}$$

where $\phi(\pi)$ is the observed frequency of the partition π of the set and the sum is over the 15 partitions. In 95% of such experiments, the value of χ^2 would lie between 5.63 and 26.12 if the 15 partitions were in fact equally probable.

1	1	1	1	50
1	1	1	2	56
1	1	2	1	46
1	1	2	2	55
1	1	2	3	50
1	2	1	1	51
1	2	1	2	52
1	2	1	3	42
1	2	2	1	52
1	2	2	2	52
1	2	2	3	38
1	2	3	1	55
1	2	3	2	48
1	2	3	3	51
1	2	3	4	52

13

Sequencing, Ranking, and Selection Algorithms in General Combinatorial Families (SELECT)

(A) INTRODUCTION

In Chapters 1–12 we discussed the generation of families of combinatorial objects, sequentially, and at random. Now we shall give a method which exploits a common feature of these and many other families. Not surprisingly, the unified method may not have the efficiency of the individually tailored special algorithms.

The basic property of the families we consider is that they can be constructed recursively. More precisely, there must be a recursive relationship which establishes a one-to-one correspondence between the set of objects of a certain order and a (disjoint) union of (multiple copies of) sets of objects of lower order. The recursive construction provides a natural way to sequence the objects, to find the rank (sequence number) of an object, to construct the object with a given rank, or to select an object uniformly at random.

In order to develop the idea it is convenient to consider the prototype, in which the combinatorial family consists of $A(n, k)$, a listing of all k-subsets of an n-set ($0 \leq k \leq n$). The recurrence

(1)
$$\binom{n}{k} = \binom{n-1}{k} + \binom{n-1}{k-1}$$

related to the cardinality $\binom{n}{k}$ of $A(n, k)$ is well known; however, we need the recurrence between the lists $A(n, k)$ themselves:

(2) $A(n, k) = A(n-1, k), A(n-1, k-1) \otimes \{n\}$

This formula expresses the fact that a listing of the k-subsets of $\{1, \ldots, n\}$ is obtained from a listing of the k-subsets of $\{1, \ldots, n-1\}$, followed by a listing of the $(k-1)$-subsets of $\{1, \ldots, n-1\}$, each of them adjoined with the element n.

With a particular $S \in A(n, k)$ we can associate a walk which visits the various A's beginning at $A(n, k)$. Recursively, having arrived at $A(\mu, \nu)$, proceed to $A(\mu-1, \nu-1)$ if $\mu \in S$, or to $A(\mu-1, \nu)$ if $\mu \notin S$. Terminate the walk at $A(0, 0)$.

For example, if $n = 7$, $k = 4$, $S = \{1, 3, 5, 6\}$, we obtain $A(7, 4) \rightarrow A(6, 4) \rightarrow A(5, 3) \rightarrow A(4, 2) \rightarrow A(3, 2) \rightarrow A(2, 1) \rightarrow A(1, 1) \rightarrow A(0, 0)$. Conversely, with a walk from $A(n, k)$ to $A(0, 0)$, which at each step goes from $A(\mu, \nu)$ to $A(\mu-1, \nu-1)$ or to $A(\mu-1, \nu)$, we can uniquely associate a set S: S is the set of values of μ where we pass from $A(\mu, \nu)$ to $A(\mu-1, \nu-1)$. The interesting observation is that the set is determined by the walk, and that it is not necessary to examine the nature of the $A(\mu, \nu)$; they could as well be lattice points (μ, ν) in the plane, or just vertices of some graph (we could in fact say that the set *is* the walk).

(B) GENERAL SETTING

In the general setting we deal with what we call a combinatorial family, which is a directed graph G on a set of vertices $V(G)$, with various properties:

(i) $V(G)$ has a partial order and a unique minimal element τ; G is locally finite: for each $v \in V(G)$ the set $\{x \in V(G) | x \leqslant v\}$ is finite.

(ii)   Each edge of G points from a vertex to one that is lower in the partial order. Every vertex v, except τ, has a strictly positive out-valence $\rho(v)$. Two vertices may be joined by more than one edge, but by only finitely many.

(iii)   For each vertex v, the set $E(v)$ of outgoing edges has been assigned a ranking, $0 \leq r_v(e) \leq \rho(v) - 1$ $(e \in E(v))$.

Every walk in G, starting from a vertex v, and respecting the orientation of the edges, will eventually end in τ; such a complete walk is called a *(combinatorial) object of order v*.

We shall show how the ranking (iii) of the edges induces a (basically lexicographic) ranking of the objects of each order v. With respect to this ranking we study four tasks. In addition, we have the job of showing how the various special cases fit into this general framework.

Task 1: Sequence Given an object in the family, construct the "next" object (e.g., which permutation of 7 letters follows (2614735)?).

Task 2: Rank Given an object ω of the combinatorial family, find the integer r such that ω is the rth member of the family, in the ordering implied by Task 1 (e.g., what is the rank of the partition $15 = 4 + 4 + 3 + 3 + 1$ in the list of all partitions of 15?).

Task 3: Unrank Given an integer r, construct the rth member of the family (e.g., which is the 322,576th 12-subset of 25 objects?).

Task 4: Random Select an object uniformly at random from the given family.

With each object ω we associate a *code word*. Let ω be a path with edge sequence $(e_1, e_2, \ldots, e_n)$ leading from v to τ; let $i(e)$, $t(e)$ denote the initial and terminal vertices of edge e; then, of course, $t(e_i) = i(e_{i+1})$, $1 \le i \le n$; denote this common value by v_i; furthermore, $v = v_0 = i(e_1)$, $t(e_n) = \tau$. Using the ranking functions for each $E(v_i)$ (see (iii)) we obtain the *code word* for τ:

$$(r_{v_0}(e_1), \ldots, r_{v_{n-1}}(e_n))$$

It is obvious that ω is completely determined by v and its code word. The sequencing of the objects of order v is now by the lexicographical ordering of the code words. Recursively, this ordering may be described as follows: Let $e_1, \ldots, e_{\rho(v)}$ be the outgoing edges from v, in order (i.e., $r_v(e_i) = i - 1$), then the ordering of the list $A(v)$ of objects of order v is implied by the concatenation of the lists in

$$(3) \qquad A(v) = e_1 \otimes A(t(e_1)), \ e_2 \otimes A(t(e_2)), \ \ldots$$

where $e \otimes A(t(e))$ denotes the list of paths whose first edge is e, followed by a path in $A(t(e))$ which begins at the endpoint of e.

To accomplish Task 1 (sequencing) in a combinatorial family, we have the simple

ALGORITHM NEXT

[Given an object ω of order v; output its immediate successor in the list of all objects of order v, ordered lexicographically by their codewords.]

Begin at τ. Back up along the walk ω starting from τ until reaching, for the first time, an edge e which is not the last outbound edge from its initial vertex. If no such e exists, terminate the algorithm. Else, replace e by the next one, say e', and complete the new walk from the final vertex of e' by choosing, at each step, edge 0 until τ is reached. ∎

Two auxiliary counting functions are needed to accomplish Task 2.

If $b(v) = |A(v)|$ is the *number* of combinatorial objects of order v, then we have, evidently, the recurrence

$$(4) \qquad b(v) = \begin{cases} \displaystyle\sum_w g_{vw} b(w) & v \in V(G) - \tau \\ 1 & v = \tau \end{cases}$$

in which g_{vw} is the number of outgoing edges from v to w. Of course, $\rho(v) = \Sigma_w \, g_{vw}$.

The second function assigns a partial sum of b's to each edge: If $e \in E(v)$, let $e' < e$ mean "$e' \in E(v)$ and $r_v(e') < r_v(e)$." We define

$$(5) \qquad f(e) = \sum_{e' < e} b(t(e'))$$

Thus $f(e)$ counts the number of walks which *precede* the *first* walk from v whose first edge is e, see (3).

Task 2, ranking, is therefore done very easily. Let ω be an object of order v, with edge sequence $(e_1, \ldots, e_n)$. Then the rank $r(\omega)$ is given by

$$r(\omega) = \sum_{i=1}^{n} f(e_i)$$

and, as ω runs over the objects of order v, $r(\omega)$ runs over the integers $0, 1, 2, \ldots, b(v) - 1$.

For Task 3, we are given an integer r and the vertex v, and we are asked to find the object of rank r. To do this, begin at v with $r' = r$. Generically, having arrived at w with r', exit along the highest outbound edge e for which $f(e) \leqq r'$, set $r' \leftarrow r' - f(e)$, and continue from $t(e)$ with r'. Halt at τ with 0.

Finally, we give two methods for Task 4, choosing a random object of order v. First, we could select an integer r at random in $[0, b(v) - 1]$, then unrank r via performing Task 3.

More in the spirit of the earlier algorithms in this book is the following alternate method, which also has superior numerical stability properties:

Begin at v. Generically, having arrived at w, choose the next vertex w' in the walk according to the probabilities

(6) $$\text{Prob}(w') = g_{ww'} b(w')/ b(w) \qquad (w' \in G)$$

Then choose one of the $g_{ww'}$ edges $w \to w'$ uniformly at random. Continue from w'. Halt at τ.

(C) EXAMPLES

As stated in the Introduction, in order for this theory to work, there must be a one-to-one correspondence between the set $A(w)$ of objects of order v, and a (disjoint) union of (multiple copies of) sets of objects $A(w)$, with $w \leqq v$. As a practical matter, in many examples the sets of some order are determined by two integer parameters; as a result, the graph G will have its vertices at lattice points in the plane; the partial order $(n, k) \leqq (n', k')$ is defined by $n \leq n'$ and $k \leq k'$. With few exceptions, the edges from (n, k) will run to $(n - 1, k)$ and $(n - 1, k - 1)$; usually, there will be more than one edge between certain pairs of vertices, reflecting the multiplicities of the copies in the recursive relationship.

Family 1 The k-subsets of an n-set. The vertices are pairs (n, k), $0 \leq k \leq n$; (n, k) is joined to $(n - 1, k)$ and $(n - 1, k - 1)$ each (provided the entries satisfy the stated inequalities) with one edge; they are numbered 0 and 1, in this order.

Family 2 The set $A(n, k)$ of partitions of a set of n elements into k classes $(0 \leq k \leq n)$. If $S(n, k) = |A(n, k)|$, then we have

$$S(n, k) = S(n - 1, k - 1) + k\, S(n - 1, k)$$

and for the list $A(n, k)$ we have

$$A(n, k) = i_{n,1} A(n - 1, k), \ i_{n,2} A(n - 1, k),$$
$$\ldots, \ i_{n,k} A(n - 1, k), \ i_{n,k} A(n - 1, k - 1)$$

where $i_{n,l}A(m, k)$ is obtained from $A(m, k)$ by inserting n into the lth class; if $l = k + 1$, the class is a new one. Vertex (n, k) is joined to $(n - 1, k)$ by k edges, numbered 0 through $k - 1$, and is joined to $(n - 1, k - 1)$ by one edge numbered k.

Family 3 The permutations of an n-set which have exactly k cycles. If $\tilde{S}(n, k)$ is the number of these, then

(7) $$\tilde{S}(n, k) = (n - 1)\tilde{S}(n - 1, k) + \tilde{S}(n - 1, k - 1)$$

Here the second term counts the permutations where n is a fixed point, and the first term counts those where n lives in a cycle with lower letters. The graph G is again on the lattice points of the plane, where now from (μ, ν) $(\mu \geq \nu + 1, \nu \geq 0)$ there go $\mu - 1$ edges, numbered $0, 1, \ldots, \mu - 2$, westbound to $(\mu - 1, \nu)$, and one edge, numbered $\mu - 1$, southwest to $(\mu - 1, \nu - 1)$. There is a 1–1 correspondence between walks from (n, k) to $(0, 0)$ and permutations of n letters with k cycles.

Family 4 Vector subspaces of dimension k of n-dimensional space over a finite field of q elements.
   Let $\begin{bmatrix} n \\ k \end{bmatrix}_q$ denote the number of such subspaces. Then we have

(8) $$\begin{bmatrix} n \\ k \end{bmatrix}_q = q^k \begin{bmatrix} n - 1 \\ k \end{bmatrix}_q + \begin{bmatrix} n - 1 \\ k - 1 \end{bmatrix}_q$$

The combinatorial meaning of this recurrence will be discussed below under "decoding." The graph G has for its vertices the lattice points of the plane, and from (μ, ν) there go q^ν westbound edges to $(\mu - 1, \nu)$, numbered $0, 1, \ldots, q^\nu - 1$ $(\mu \geq \nu + 1, \nu \geq 0)$ and one southwestbound edge, numbered q^ν $(\mu \geq \nu \geq 0)$. There is a 1–1 correspondence between vector subspaces of dimension k of n-dimensional space over $GF(q)$, on the one hand, and the set of all walks from (n, k) to $(0, 0)$ on the graph G. The correspondence will be explicitly described below.

Family 5 Permutations of n letters with k runs.
   By a "run" in a permutation we mean a maximal ascending consecutive subsequence, e.g., the permutation

(9) (82157346)

of 8 letters has four runs.
   If the Eulerian number $\langle {n \atop k} \rangle$ is the number of such permutations then

$$(10) \qquad \left\langle {n \atop k} \right\rangle = k \left\langle {n-1 \atop k} \right\rangle + (n-k+1) \left\langle {n-1 \atop k-1} \right\rangle.$$

Indeed, from each of the $\left\langle {n-1 \atop k} \right\rangle$ permutations of $n-1$ letters with k runs, we can make k permutations with k runs by inserting the letter n at the end of one of the runs. Finally, from each of the $\left\langle {n-1 \atop k-1} \right\rangle$ permutations of $n-1$ letters with $k-1$ runs, we can make $n-k+1$ permutations with k runs by inserting n interior to one of the runs.

Our graph G again has for its vertices the lattice points of the plane. From the vertex (μ, ν) $(\mu \geqq \nu + 1, \nu \geqq 1)$ there go ν westbound edges, numbered $0, \ldots, \nu - 1$, to $(\mu - 1, \nu)$. From (μ, ν) $(\mu \geqq \nu, \nu > 1)$ there go $\mu - \nu + 1$ southwestbound edges, numbered $\nu, \ldots, \mu$, to $(\mu - 1, \nu - 1)$. From $(1, 1)$ one edge, numbered 0, goes to $(0, 1)$, the terminal vertex.

Family 6 Partitions of the integer n whose largest part is k.

If $p(n, k)$ is the number of these, then Eq. (4) of Chapter 10 is the recurrence

$$(11) \qquad p(n, k) = p(n-k, k) + p(n-1, k-1)$$

The second term counts those partitions whose largest part is less than k, the first term counts those whose largest part is equal to k.

On the lattice points of the plane once more, from (μ, ν) $(\mu \geqq 2\nu \geqq 0)$ there goes westbound to $(\mu - \nu, \nu)$ a single edge, numbered 0, and from (μ, ν) $(\mu \geqq \nu \geqq 2$ or $\mu = \nu = 1)$ there goes a single edge to $(\mu - 1, \nu - 1)$, numbered 1 unless $\mu = \nu = 1$, and numbered 0 in that case. The graph is atypical in that a westbound step can be more than one unit long.

Family 7 Compositions of n into k parts.

We have seen in Chapter 5 that there are

$$(12) \qquad b(n, k) = \binom{n+k-1}{n}$$

of these compositions, and so the recurrence

$$(13) \qquad b(n, k) = b(n, k-1) + b(n-1, k)$$

holds. In (13), the first term on the right counts those compositions whose first part is 0, the second term counts the others.

Now we have a graph on the lattice points of the plane where from the point (μ, ν) $(\mu \geqq 1, \nu \geqq 1)$ there goes one edge to $(\mu - 1, \nu)$ and from (μ, ν) $(\mu \geqq 0, \nu \geqq 2)$ there goes one edge to $(\mu, \nu - 1)$. The graph is atypical in that the "southwest" edge here goes south. As usual, there is a 1–1 correspondence between paths from (n, k) to $(0, 1)$ and compositions of n into k parts.

(D) THE FORMAL ALGORITHMS

In this section we give algorithms for performing any of the four tasks (sequencing, ranking, unranking, random selection) on any of the seven combinatorial families mentioned in the previous section.

The algorithms of this section will apply to the *coded form* of the combinatorial objects, that is, we suppose that the objects are represented as *paths* on their appropriate *graphs*. For some ordinary applications this is sufficient. For most applications it will be necessary to translate the coded form of the object to one of the familiar forms which are more easily recognizable. This process, called *decoding*, is discussed in Section 6 of this chapter. For some applications it may be necessary to translate a familiar form of an object into its coded form (*encoding*). We leave these encoding algorithms to the reader, who will, we think, find them to be easy inversions of the corresponding decoding algorithms.

The coded form of a combinatorial object, as we have said, is that of a path from a vertex v to the terminal vertex τ in a graph G with numbered edges. We specialize at once to the case where the vertices of the graph are lattice points (μ, ν) of the plane, as is the case in the seven families so far mentioned. Then an object of order (n, k) is represented by a vertex-sequence and edge-number sequence of the precise form

$$(14) \quad (n, k) = (\mu_1, \nu_1) \xrightarrow{\text{edge}_1} (\mu_2, \nu_2) \xrightarrow{\text{edge}_2} \cdot \cdot \cdot \xrightarrow{\text{edge}_{m-1}} (\mu_m, \nu_m) \xrightarrow{\text{edge}_m} \tau.$$

Thus $(\mu_1, \nu_1) = (n, k)$ is the *order* of the object, and edge(i) is the number of the edge of the path which goes from (μ_i, ν_i) to (μ_{i+1}, ν_{i+1}). There is evidently redundant information here, since if we are given (n, k) and the edges we can deduce the vertex sequence. Nonetheless, it is convenient to have all of these arrays.

Next, we examine the seven recurrence relations (1), (6), (7), (8), (10), (11), and (13) of the families in question. We observe that they are all of the form

$$(15) \qquad b(\mu, \nu) = \varphi(\mu, \nu)b(\mu_w, \nu) + \psi(\mu, \nu)b(\mu_s, \nu - 1)$$

in which φ, ψ are certain given functions in each family, μ_w ("western" μ) is $\mu - 1$ except in Family 6, where it is $\mu - \nu$, and μ_s ("southwestern" μ) is $\mu - 1$ except in Family 7, where it is μ.

The precise form of the function φ, ψ in each of the seven families is stated in the subprograms PHI(MU,NU,FAMILY),PSI(MU,NU, FAMILY), to which the reader's attention is directed. The subprogram XNEW(M1,M2,MF,MGO) (q.v.) calculates the value of μ_w (if MGO=1) or of μ_s (if MGO=2) in the family MF, given the point $(\mu, \nu) = (M1, M2)$.

The edges outbound from the point (μ, ν) are numbered consecutively from 0 to $\varphi(\mu, \nu) + \psi(\mu, \nu) - 1$ in the counterclockwise direction. That is, the $\varphi(\mu, \nu)$ westbound edges if any, are numbered 0, 1, . . . , $\varphi(\mu, \nu) - 1$, and $\psi(\mu, \nu)$ southwestbound edges are numbered $\varphi(\mu, \nu)$, . . . , $\varphi(\mu, \nu) + \psi(\mu, \nu) - 1$. The terminal vertex τ has the property that $\varphi(\tau) + \psi(\tau) = 0$, and so all walks are followed until such a τ is encountered. Thus there is no need to tabulate the terminal vertices separately (the reader may wish to reexamine subprograms PHI,PSI and to tabulate for his own reference the seven terminal vertices of the families considered).

The algorithm SELECT, which follows, is the universal portion of the mechanism for performing the four tasks on the seven families. It assumes that the counting numbers $b(n, k)$ (=number of objects of order (n, k)) have been precomputed.

ALGORITHM SELECT

(A)   *[Entry for Task 4]* $r \leftarrow b(n, k) * \xi$.

(B)   *[Entry for Task 3]* $j \leftarrow 1; r' \leftarrow r$.

(C)   *[Extend path from jth step]* $(\mu_j, \nu_j) \leftarrow (n, k); m \leftarrow j$.

(D)   *[Reached terminal vertex?]*   If $\varphi_m + \psi_m = 0$, set $m \leftarrow m - 1$ and exit; *[southwest edge?]* if $r' \geqq \varphi_m b_w(m)$, to (E); *[next edge goes west]* edge$(m) \leftarrow \lfloor r'/b_w(m) \rfloor$; *[decrease remaining rank]* $r' \leftarrow r' - $ edge$(m)b_w(m)$; $(\mu_{m+1}, \nu_{m+1}) \leftarrow (x_w(m), \nu_m)$; $m \leftarrow m + 1$; to (D).

(E)   *[Next edge goes southwest]* $r' \leftarrow r' - b_w(m)\varphi_m$; edge$(m) \leftarrow \varphi_m + \lfloor r'/b_s(m) \rfloor$; *[decrease remaining rank]* $r' \leftarrow r' - ($edge$(m) - \varphi_m)b_s(m)$; $(\mu_{m+1}, \nu_{m+1}) \leftarrow (x_s(m), \nu_m - 1)$; $m \leftarrow m + 1$; *to* (D).

(F)   *[Entry for Task 2]* $r \leftarrow 0$
    For $j = 1, m - 1$ do:
    (F1) If $\nu_{j+1} \neq \nu_j$, to (F2); $r \leftarrow r + $ edge$(j)b_w(j)$; next j.
    (F2) $r \leftarrow r + \varphi_j b_w(j) + ($edge$(j) - \varphi_j)b_s(j)$; next j.
    End
    Exit

(G) [*First entry for Task 1*] $r \leftarrow 0$; to **(B)**.

(H) [*All later entries for Task 1*] $j \leftarrow m$.

(K) [*Is jth edge moveable?*] If edge(j) $< \varphi_j + \psi_j - 1$, to **(L)**; [*Backtrack*] $j \leftarrow j - 1$; If $j \neq 0$, to **(K)**; Final exit.

(L) [*Augment edge*] edge(j) $\leftarrow$ edge(j) $+ 1$; $l \leftarrow j + 1$; If edge(j) $\neq \varphi_j$, to **(M)**; $(\mu_1, \nu_l) \leftarrow (x^s(l - 1), \nu_l - 1)$.

(M) [*Extend with 0's to terminal vertex*] $r' \leftarrow 0$; $m \leftarrow l$; to **(D)** ∎

The following abbreviations were used in the statement of the algorithm.

r The rank of an object.

ξ A random number.

$b(n, k)$ The number of objects of order (n, k).

φ_m $\varphi(\mu_m, \nu_m)$.

ψ_m $\psi(\mu_m, \nu_m)$.

$b_w(m)$ The value of $b(\mu', \nu')$ at the point west of (μ_m, ν_m).

$b_s(m)$ The value of $b(\mu', \nu')$ at the point southwest of (μ_w, ν_m).

$x_w(m)$ The x-coordinate of the point west of (μ_m, ν_m).

$x_s(m)$ The x-coordinate of the point southwest of (μ_m, ν_m).

SUBROUTINE SPECIFICATIONS

We depart slightly from the format of the previous chapters here in order to expand somewhat the description of the subroutine variables.

The calling statement is

```
CALL SELECT(FAMILY,TASK,N,K,MU,NU,EDGE,M,
                              NEWONE,RANK,B)
```

in which

(a) FAMILY is an integer variable, $1 \leq$ FAMILY ≤ 7, which describes the combinatorial family in question, according to the scheme

FAMILY = 1   K-subsets of an N-set

$\quad\quad\quad = 2$ partitions of N objects into K classes

$\quad\quad\quad = 3$ permutations of N objects with K cycles

$\quad\quad\quad = 4$ vector subspaces of dimension K of N-dimensional space over $GF(Q)$‡

$\quad\quad\quad = 5$ permutations of N letters with K runs

‡ Q is set to 2 by a DATA statement in the FUNCTION PH1 subprogram. It can easily be changed by the user, if desired.

$= 6$ partitions of N whose largest part is K
$= 7$ compositions of N into K parts.

(b)   TASK is an integer variable, $1 \leq$ TASK ≤ 4, which controls the functioning of the routine as follows.

TASK $= 1$: Sequencing. The subroutine produces the next object of the FAMILY and exits. Required input data are FAMILY, TASK= 1, N, K, as well as the description of the previous object MU, NU, EDGE, M. The logical variable NEWONE is set by the user to . FALSE. to initiate a new sequence. The subroutine returns NEWONE= . TRUE. with each output object, including the last. When the subroutine is called once more, after the last object has been delivered, the value NEWONE= . FALSE. will be returned. Thus NEWONE is similar in spirit to MTC of Chapters 1, 3, 5, 7, 9, 11, but not identical.

TASK $= 2$: Ranking. Given an object, the subroutine computes its rank among those of its order, and exits. Input data are FAMILY; TASK=2; the order (N, K); the object MU, NU, EDGE, M. Output is RANK.

TASK $= 3$: Unranking. The subroutine computes the object of given rank. Input data are FAMILY; TASK=3; the order (N, K); the RANK. Output is the object MU, NU, EDGE, M.

TASK $= 4$: Random selection. The subroutine chooses, uniformly at random, an object of given order. Input data are FAMILY; TASK=4; the order (N, K). Output is the object MU, NU, EDGE, M.

(c)   N and K are integers which give the order of the objects considered.

(d)   MU, NU, EDGE are integer arrays of length M. These describe an object precisely as given by the form (14).

(e)   M is the length of the arrays MU, NU, EDGE, in the sense of (14).

(f)   NEWONE is a logical variable whose purpose is described under TASK=1 in (b) above.

(g)   RANK is an integer variable. It is the rank of the object under consideration among those of its order (N, K), where $0 \leq$ RANK $\leq b(n, k) - 1$.

(h)   B is an N×K integer array, where B(I, J) is the number of objects of order I, J. These are computed automatically, as needed, by the subroutine, and may be ignored by the user. On each call to the subroutine, the input order N, K is checked against NLAST, KLAST. If either N>NLAST or K>KLAST, the values B(I, J) are computed and stored for all $1 \leq$ I$\leq$N, $1 \leq$ J$\leq$K, and then (NLAST, KLAST) is set to (N, K). Otherwise, no values are computed.

There are 118 instructions in the subroutine.

```
      SUBROUTINE SELECT(FAMILY,TASK,N,K,MU,NU,EDGE,M,
      NEWONE,RANK,B)
      IMPLICIT INTEGER(A-Z)
      LOGICAL NEWONE
      REAL RAND
      INTEGER MU(N),NU(N),EDGE(N),B(10,10)
      DATA NLAST,KLAST,FLAST/0,0,0/
      IF(N.LE.NLAST.AND.K.LE.KLAST.AND.FAMILY.EQ.
    * FLAST) GO TO 205
      NLAST=N
      KLAST=K
      FLAST=FAMILY
      DO 201  M1=1,N
      DO 201  K1=1,K
201   B(M1,K1)=PHI(M1,K1,FAMILY)*BNEW(M1,K1,FAMILY,
    * 1,B)+PSI(M1,K1,FAMILY
    * )*BNEW(M1,K1,FAMILY,2,B)
205   GO TO (100,400,200,300),TASK
300   RANK=RAND(1)*B(N,K)
200   J=1
      RANKP=RANK
500   MU(J)=N
      NU(J)=K
      M=J
510   T1=PHI(MU(M),NU(M),FAMILY)
      IF(T1+PSI(MU(M),NU(M),FAMILY).NE.0) GO TO 518
      M=M-1
      RETURN
518   B1=BNEW(MU(M),NU(M),FAMILY,1,B)
      IF(RANKP.GE.T1*B1) GO TO 520
512   EDGE(M)=RANKP/B1
      RANKP=RANKP-EDGE(M)*B1
      MU(M+1)=XNEW(MU(M),NU(M),FAMILY,1)
      NU(M+1)=NU(M)
515   M=M+1
      GO TO 510
520   RANKP=RANKP-B1*T1
      B2=BNEW(MU(M),NU(M),FAMILY,2,B)
      EDGE(M)=T1+RANKP/B2
      RANKP=RANKP-(EDGE(M)-T1)*B2
      MU(M+1)=XNEW(MU(M),NU(M),FAMILY,2)
      NU(M+1)=NU(M)-1
```

```
       GO TO 515
400    RANK=0
       M1=M-1
       DO 401  J=1,M1
       IF(NU(J+1).NE.NU(J)) GO TO 402
       RANK=RANK+EDGE(J)*BNEW(MU(J),NU(J),
     * FAMILY,1,B)
       GO TO 401
402    RANK=RANK+PHI(MU(J),NU(J),FAMILY)*BNEW
     * (MU(J),NU(J),FAMILY,1,B)
     * +(EDGE(J)-PHI(MU(J),NU(J),FAMILY))*BNEW
     * (MU(J),NU(J),FAMILY,2,B)
401    CONTINUE
       RETURN
100    IF(NEWONE) GO TO 105
       NEWONE=.TRUE.
       RANK=0
       GO TO 200
105    J=M
120    T=PHI(MU(J),NU(J),FAMILY)
       IF(EDGE(J).LT.T+PSI(MU(J),NU(J),FAMILY)-1)
     * GO TO 130
       J=J-1
       IF(J.NE.0) GO TO 120
       NEWONE=.FALSE.
       RETURN
130    EDGE(J)=EDGE(J)+1
       L=J+1
       IF(EDGE(J).NE.T) GO TO 140
       NU(L)=NU(L)-1
       MU(L)=XNEW(MU(L-1),NU(L-1),FAMILY,2)
140    RANKP=0
       M=L
       GO TO 510
       END

       INTEGER FUNCTION XNEW(M1,M2,MF,MGO)
       XNEW=M1-1
       IF(MF.EQ.6.AND.MGO.EQ.1) XNEW=M1-M2
       IF(MF.EQ.7.AND.MGO.EQ.2) XNEW=M1
       RETURN
       END
```

```
      INTEGER FUNCTION BNEW(M1,M2,MF,MGO,B)
      INTEGER B(10,10),XNEW
      BNEW=1
      LX=XNEW(M1,M2,MF,MGO)
      LY=M2-MGO+1
      IF(LX*LY.GT.0) BNEW=B(LX,LY)
      RETURN
      END


      FUNCTION PHI(MU,NU,FAMILY)
      IMPLICIT INTEGER (A-Z)
      DATA Q/2/
      PHI=0
      GO TO (10,20,30,40,50,60,70),FAMILY
10    IF(MU.GE.NU+1.AND.NU.GE.0) PHI=1
      RETURN
20    IF(MU.GE.NU+1.AND.NU.GE.1) PHI=NU
      RETURN
30    IF(MU.GE.NU+1.AND.NU.GE.0) PHI=MU-1
      RETURN
40    IF(MU.GE.NU+1.AND.NU.GE.0) PHI=Q**NU
      RETURN
50    IF(MU.GE.NU+1.AND.NU.GE.1) PHI=NU
      IF(MU.EQ.1.AND.NU.EQ.1) PHI=1
      RETURN
60    IF(MU.GE.2*NU.AND.NU.GE.1) PHI=1
      RETURN
70    IF(MU.GE.1.AND.NU.GE.1) PHI=1
      RETURN
      END


      FUNCTION PSI(MU,NU,FAMILY)
      IMPLICIT INTEGER(A-Z)
      PSI=0
      GO TO (10,20,20,10,30,40,50),FAMILY
10    IF(MU.GE.NU.AND.NU.GE.1) PSI=1
      RETURN
20    IF((MU.GE.NU.AND.NU.GE.2).OR.
     *  (MU.EQ.1.AND.NU.EQ.1)) PSI=1
      RETURN
30    IF(MU.GE.NU.AND.NU.GE.2) PSI=MU-NU+1
```

```
       RETURN
40     IF((MU.GE.NU.AND.NU.GE.2).OR.
  *    (MU.EQ.1.AND.NU.EQ.1)) PSI=1
       RETURN
50     IF(MU.GE.0.AND.NU.GE.2) PSI=1
       RETURN
       END
```

(E) DECODING

The decoding problem is that of translating the representation of the object as a path

$$(16) \qquad (n, k) = (\mu_1, \nu_1) \xrightarrow{\text{edge}_1} (\mu_2, \nu_2) \xrightarrow{\text{edge}_2} \cdots \xrightarrow{\text{edge}_{m-1}} (\mu_m, \nu_m) \xrightarrow{\text{edge}_m} \tau$$

into a representation in a familiar form. In all cases, this is done by considering the combinatorial meaning of the recurrence relation of the family.

In general, we can begin at the terminal vertex with a suitable terminal object, and back up along the path. At a generic step, suppose we go from (μ, ν) to (μ', ν') via edge e. If $\nu' = \nu$ we have an eastbound step, and we perform operation E (described below). Otherwise we have a northeastbound step, and we perform operation N (described below). When (n, k) is reached, we exit with the familiar form of the object.

Family 1 k-subsets of an n-set.

The set S is initially empty. In operation E we do nothing. In operation N we adjoin element μ' to the set S. When (n, k) is reached, the set S is the desired output set in familiar form.

Family 2 Partitions of n objects into k classes.

The partiton P is initially empty. In operation E we insert the letter μ' into class number e of the partition P (the classes are numbered 0, 1, . . .). In operation N we adjoin the letter μ' as a new singleton class to P.

The reader may wish to verify that

$$(5, 3) \xrightarrow{1} (4, 3) \xrightarrow{3} (3, 2) \to (2, 2) \xrightarrow{0} (1, 1) \xrightarrow{0} (0, 0)$$

decodes into the partition of 5 elements into 3 classes (1) (235) (4) when written in familiar form.

Family 3 Permutations of n objects with k cycles.

The permutation P is initially empty. In operation E we must insert the letter μ' into the eth "space" in the permutation P, where "space" means the following: μ letters have already been inserted into the permutation P. Think of the cycles of P as necklaces. There are μ spaces between consecutive "beads." Number these $0, 1, \ldots$ in some standard way. Then insert μ' into the space number e.

In operation N we adjoin the letter μ' as a singleton cycle to the permutation P.

Family 4 k-subspaces of n-spaces over $GF(q)$.

The "familiar form" of a subspace is taken to be a $k \times n$ matrix V over $GF(q)$ whose rows are a basis. Let the matrix V be initially 0×0. In operation E we will adjoin to V a new column of ν field elements, namely the column of the digits of the number e when written as a q-ary number.

In operation N we will border the matrix V with a new row and column: a new row of zeros, a new column of zeros, and a single 1 at their intersection (see [CW1] for details).

Family 5 Permutations of n letters with k runs.

The terminal vertex is $(0, 1)$. Initially we have the empty permutation Π. On an operation E we insert μ' at the end of the eth run of the permutation Π (runs numbered $0, 1, \ldots$).

In operation N the letter μ' is inserted into the eth one of the spaces interior to existing runs (also numbered $0, 1, \ldots$).

Family 6 Partitions of n whose largest part is k.

We begin with the empty "partition" $\Pi: 0 = 0$. On each operation E we replicate the largest part of Π. On each operation N we add 1 to the largest part of Π.

Family 7 Compositions of n into k parts.

Begin with the empty composition κ. An operation E adds 1 to the first part of the composition κ. Each operation N adjoins a new first part to κ, namely 0.

While these algorithms describe the decoding process, of course they need not be, in each case, the most efficient decoding method. A better way to decode permutations of n letters with k cycles, for example, is to begin at (n, k) and follow the path forwards, with an initially blank array $a_1, \ldots, a_n$. At the jth step of the walk, if the number of cycles not yet started is equal to the number of edges not yet traversed, fill all remaining blanks in a with $-1, -2, -3, \ldots$ and exit. Else, if $edge(j) = n - j$, insert $-(n + 1 - j)$ into the $(e_j + 1)$th blank space on the array a, while if $e_j < n - j$, put $n + 1 - j$ into the $(e_j + 2)$th blank space. At termination, the array $a_1, \ldots, a_n$ holds the usual cycle form of the permutation, with the leftmost element of each cycle flagged with a negative sign.

SAMPLE OUTPUT

With $k = 3$, $n = 5$, the three tables below show the first ten objects of order (5, 3) in Families 1, 2, 3. For each object we display its RANK, its EDGE, MU, NU arrays, and its familiar form

Family 1   3-subsets of a 5-set.

RANK	EDGE	MU	NU	SUBSET
0	00000	543210	333210	{1,2,3}
1	01000	543210	333221	{1,2,4}
2	01101	543210	333211	{1,3,4}
3	01110	543210	333210	{2,3,4}
4	10000	543210	332221	{1,2,5}
5	10100	543210	332211	{1,3,5}
6	10110	543210	332210	{2,3,5}
7	11000	543210	332111	{1,4,5}
8	11010	543210	332110	{2,4,5}
9	11100	543210	332100	{3,4,5}

Family 2   Partitions of 1, 2, 3, 4, 5 into three classes.

RANK	EDGE	MU	NU	PARTITION
0	00000	543210	333210	(145)(2)(3)
1	01000	543210	333210	(15)(24)(3)
2	02000	543210	333210	(15)(2)(34)
3	03000	543210	332210	(135)(2)(4)
4	03100	543210	332210	(15)(23)(4)
5	03200	543210	332110	(125)(3)(4)
6	10000	543210	333210	(14)(25)(3)
7	11000	543210	333210	(1)(245)(3)
8	12000	543210	333210	1)(25)(34)
9	13000	543210	332210	(13)(25)(4)

Family 3 Permutations of 12345 with 3 cycles.

RANK	EDGE	MU	NU	PERMUTATION
0	00000	543210	333210	(154)(2)(3)
1	01000	543210	333210	(15)(24)(3)
2	02000	543210	333210	(15)(2)(34)
3	03000	543210	332210	(153)(2)(4)
4	03100	543210	332210	(15)(23)(4)
5	03200	543210	332110	(152)(3)(4)
6	10000	543210	333210	(145)(2)(3)
7	11000	543210	333210	(1)(254)(3)
8	12000	543210	333210	(1)(25)(34)
9	13000	543210	332210	(135)(2)(4)

14

Young Tableaux (NEXYTB/RANYTB)

(A) INTRODUCTION

Consider the partition

(1) $$\pi: 6 = 3 + 2 + 1$$

of $n = 6$. With the partition π we associate a *shape*, namely

(2)

in which the successive parts of the partition are the lengths of successive rows of the shape.

Next we fill in the 6 squares in the shape with the letters 1, . . . , 6 subject to the condition that the integers in every row or column form an increasing sequence. Thus

(3)

1	3	4
2	6	
5		

1	2	6
3	4	
5		

1	2	6
3	5	
4		

are three different ways of doing this. The filled-in shape in each case is called a Young Tableau of shape π.

For the partition π shown above there are in fact 16 tableaux which have its shape.

For the partition

(4) $$\pi: 8 = 3 + 3 + 1 + 1$$

whose shape is

(5)
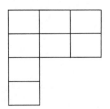

it turns out that there are exactly 56 Young tableaux.

We now describe the general formula which gives the number of tableaux of shape π. Associated with each of the n squares in a tableau is an integer called the "hook length" at that square. To find the hook length we count the number of squares to the right of the given one, in the same row, add the number below the given one, in the same column, and add 1 to count the given square itself.

For example, for the shape (5) we show below each hook length written into its square.

(6)

6	3	2
5	2	1
2		
1		

Let π be a partition of the integer n, then the number of Young tableaux of the given shape π is

(7) $$N(\pi) = \frac{n!}{\text{(product of all hook lengths)}}$$

where n is the integer of which π is a partition. Thus from (6) the number of tableaux belonging to the shape (4) is

$$\frac{8!}{1 \cdot 1 \cdot 2 \cdot 2 \cdot 2 \cdot 3 \cdot 5 \cdot 6} = 56$$

It is a remarkable fact that the algorithm which we will give in this chapter for selecting a Young tableau uniformly at random (u.a.r.) from among those of a given shape also provides a proof of the counting theorem (7), a proof, furthermore, in which the role of the hooks is a natural one. The algorithm and the proof of the counting theorem are due to Greene, Nijenhuis, and Wilf [GNW1].

Next we note that the highest letter n in a tableau must appear at the end of a row and of a column. Let us call such a position a *corner* of the shape. There are as many corners as there are *distinct* parts in the partition π.

We identify the Young tableaux with the general theory of combinatorial families described in the preceding chapter. Consider a graph G which has a vertex π corresponding to each partition of each positive integer n, and the "empty partition" $0 = 0$.

We draw a directed edge from a vertex π to a vertex π' in G if the shape of π' is obtained from the shape of π by deleting exactly one corner.

Thus the out-valence of π is equal to the number of distinct parts of π.

Now imagine a walk ω which begins at π, and follows directed edges of G to the terminal vertex

$$\tau: 0 = 0$$

Then ω is a Young tableau of shape π. Indeed, we begin with the shape π and walk along the path ω. As we traverse each edge we insert the highest letter not yet inserted into the corner whose deletion corresponds to the edge.

Consider, for example, the following walk:

(8)

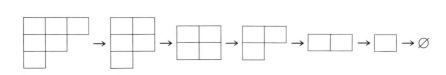

On the first edge we insert 6 into the corner which is deleted, then 5 into the next corner, . . . , obtaining the tableau

(9)

1	2	6
3	4	
5		

We see that the walk is a recursive recipe for filling in the empty shape.

The family of Young tableaux is thereby identified as a family of walks on a graph, in which the shape plays the part of the "order." The difference between this family and the ones which are treated by the program of the preceding chapter is that the vertex set here is not, in a natural way, a set of lattice points in the first quadrant of the plane.

Hence, in this chapter we describe separate algorithms for sequencing and random selection of Young tableaux.

(B) LEXICOGRAPHIC SEQUENCING

First we consider passing from a given tableau T to its immediate lexicographic successor.

Observe that the "first" tableau of shape π is obtained by inserting n into the first (topmost) corner, then $n - 1$ into the first corner of the remaining shape, etc. The first tableau of shape (1) is

(10)

1	4	6
2	5	
3		

while the last tableau of that shape is

(11)

1	2	3
4	5	
6		

Thus in the first one we write the letters $1, \ldots, n$ down the columns consecutively beginning on column 1, while in the last tableau the letters are written consecutively across the rows, top to bottom.

Now to find the successor of a given T, following algorithm NEXT of the previous chapter, we see that we must

(a) locate the smallest integer j in T which does not occupy the bottom corner of the subtableau T_j formed by $1, 2, \ldots, j$

(b) move the letter j to the next lower corner position in T_j.

(c) replace all other entries in the shape T_j by entering the numbers $1, 2, \ldots, j-1$ consecutively down the columns, beginning with column 1.

For example, let

(12)
$$T = $$

1	2	6	9
3	4	7	
5	10	11	
8	12		

The integer j in step (a) is 6 because in its subtableau

(13)
$$T_6 = $$

1	2	6
3	4	
5		

it does not occupy the last corner position, and it is the least such integer.

Following step (b) we move 6 into the next corner of T_6, thus

(14)
$$T_6^{(new)} = $$

	6	

After step (c), the subtableau has become

(15)
$$T_6^{(new)} = $$

1	4	5
2	6	
3		

The successor of the full tableau has the new T_6 embedded in the otherwise unchanged T as

(16)
$$T^{(\text{next})} = \begin{array}{|c|c|c|c|} \hline 1 & 4 & 5 & 9 \\ \hline 2 & 6 & 7 \\ \cline{1-3} 3 & 10 & 11 \\ \cline{1-3} 8 & 12 \\ \cline{1-2} \end{array}$$

It is convenient to represent a tableau in a computer by means of its Y-vector, which is defined by

(17) $y(i) = $ the number of the row which contains i $(i = 1, n)$

Thus the vector associated with the tableau T of (12) is

(18) $\qquad Y = (1, 1, 2, 2, 3, 1, 2, 4, 1, 3, 3, 4)$

while that of the successor (16) of T is

(19) $\qquad Y^{(\text{next})} = (1, 2, 3, 1, 1, 2, 2, 4, 1, 3, 3, 4)$

The reader should satisfy himself that the Y-vector of a tableau uniquely describes the tableau.

From the Y-vector of T it is easy to locate the letter j which we need in step (a) of the sequencing algorithm: it is the least j such that $y(j) < y(j - 1)$.

The complete sequencing algorithm follows.

ALGORITHM NEXYTB

[Enter at step (**A**) with the Y-vector $y(1), \ldots , y(n)$ of a tableau of order n and exit with the Y-vector of its successor, if it has one, or at "final exit," if it has none.

Enter at step (**C**) with a partition $n = \lambda(1) + \cdots + \lambda(n)$ in which the $\lambda(i)$ decrease, and a zero entry is inserted in the array λ after the last part of the partition, and exit with the first y-vector of shape λ.]

(**A**) If $n = 1$, final exit;
  $\lambda(1) \leftarrow 1; \lambda(i) \leftarrow 0 \ (i = 2, n)$ [The array λ will now be set to the shape of the subtableau of the letters $[1, 2, \ldots , j]$, where j is the point of decrease of Y]
    For $j = 2, n$ do: $\{\lambda(y_j) \leftarrow \lambda(y_j) + 1;$ *If* $y_j < y_{j-1}$, *to* (**B**)$\}$
    Final exit.
(**B**) [*Find new row for letter j*] $t \leftarrow \lambda(1 + y_j); i \leftarrow k;$
    While $\lambda(i) \neq t$ *do:* $\{i \leftarrow i - 1\}$
    [*Move j to row i*] $y_j \leftarrow i; \lambda(i) \leftarrow \lambda(i) - 1;$ *To* (**D**).

(C) $j \leftarrow n + 1$
(D) [*Fill entries of the Y-vector corresponding to shape* λ]
    (D1) $t \leftarrow j - 1; l \leftarrow 1$.
    (D2) $r \leftarrow 1$.
    (D3) If $\lambda(r) = 0$, to (D4); $y_l \leftarrow r$; $\lambda(r) \leftarrow \lambda(r) - 1$; $l \leftarrow l + 1$;
        $r \leftarrow r + 1$; To (D3)
    (D4) If $l \leq t$, to (D2); Exit ∎

The program for SUBROUTINE NEXYTB has been written so as to facilitate its use for either of two purposes: (a) listing all tableaux of a given shape or (b) finding the successor of a given tableau.

If the routine is called with MTC= . FALSE. it interprets the input array LAMBDA as the set of parts of a partition of N ($n = \lambda_1 + \lambda_2 + \cdots + \lambda_k$), in decreasing order. On output the array Y holds the Y-vector of the lexicographically first tableau of the given shape.

If called with MTC= . TRUE. the array LAMBDA is used only for working storage, and its input contents are of no significance. In this case the Y array, on output, holds the immediate successor of the tableau that was described by the input Y array.

On output MTC is normally set to . TRUE. but is . FALSE. if the output tableau is the last of its shape. Thus, to process all tableaux of given shape we would have a main program of the form

```
    MTC= .FALSE.
    [Set array LAMBDA to desired shape; if LAMBDA has K < N non-
    zero parts, enter LAMBDA(K+1)=0]
10  CALL NEXYTB( ... )
    [Process output tableau]
    IF(MTC) GO TO 10
20  [All tableaux of this shape are done]
    .
    .
    .
```

(C) RANDOM SELECTION

Next we consider the selection uniformly at random of a Young tableau of given shape. This will be done by inserting the letter n into a corner position of the given shape with the right probability, then inserting $n - 1$ into a corner of the remaining shape, etc.

Hence there are two questions to answer: (a) What is the probability that the highest letter n lies in a given corner κ of the shape π? (b) How can we arrange to insert n into a corner κ with the correct probability?

The answer to (a) is quite simple. To question (b) there is first the obvious answer, and then the algorithm that we have chosen to program, which is a little "game" in which the letter n is moved around the "board" until it hits a corner, then stops, and if the rules of the game are observed, the probabilities turn out to be the correct ones.

First, as regards question (a), the answer is just that

$$(20) \qquad \mathcal{P}(\kappa) = \frac{N(\pi - \kappa)}{N(\pi)}$$

in which $N(\pi)$ is the number of tableaux of shape π, and $\pi - \kappa$ is the shape obtained from π by erasing the corner square κ. If we substitute (7) into (20) we obtain

$$(21) \qquad \mathcal{P}(\kappa) = \frac{\text{Product of hook lengths of } \pi}{n \cdot \text{Product of hook lengths of } \pi - \kappa}$$

In (21), the hook lengths in the numerator cancel against those in the denominator except for those in the row and column which contains κ. We can rewrite (21) in the form

$$(22) \qquad \mathcal{P}(\kappa) = \frac{1}{n} \prod{}' \left(\frac{h}{h - 1} \right)$$

where the product extends over those squares of the given shape which lie in the row or column of κ, excluding κ itself, and in which h is the hook lengths of the square.

For example, if we refer back to (6), the probability of the corner in the (2, 3) position is

$$\mathcal{P}(\kappa_1) = \frac{1}{8} \left(\frac{5}{5 - 1} \right)\left(\frac{2}{2 - 1} \right)\left(\frac{2}{2 - 1} \right) = \frac{5}{8}$$

and of the corner in the (4, 1) position is

$$\mathcal{P}(\kappa_2) = \frac{1}{8} \left(\frac{6}{6 - 1} \right)\left(\frac{5}{5 - 1} \right)\left(\frac{2}{2 - 1} \right) = \frac{3}{8}.$$

In view of the known probabilities (22) of the corners, it would of course be easy to program the selection of a corner to hold the letter n. Instead of this, however, we describe a game whose outcome locates the letter n in the desired fashion, with no computations of probabilities being required.

The proof of the validity of the game will establish the corner probability (22) independently of the counting formula (7), and then by recognizing that the corner probabilities must add up to 1, we will obtain a proof of the counting formula itself. We will have then another example of how algorithmic motivation can generate valuable points of view in pure mathematics.

Consider a chessboard of shape π on which there moves a single "one-way-rook," i.e., a piece which can move any number of spaces to the right or any number down.

Player 1 places it on any square. Players 2 and 1 then move alternately until the rook reaches a corner of the shape (board); the last mover wins.

This game is easy to analyze, and is similar to "Corner the Queen" [see *Scientific American* **236**, 134 (March 1977)]. We might call the present game "Corner the Rook." A gambler's version of "Corner the Rook" is even more germane: (bets are placed on the corners before play begins) the rook is placed on the shape in any square, chosen u.a.r.; it is then moved repeatedly, each move being made to a square which is chosen u.a.r. from among those which are available destinations.

First, as to the speed of the game, let A denote the area (number of squares) to the south and east of the first square (i, j), and let A' denote the corresponding area after one move is made. The expected value of A' is $\leq \frac{1}{2}A$, and equality holds if and only if the shape to the south and east is a rectangle. Thus only about $O(\log n)$ moves are needed before a corner is reached.

In order to analyze the game, consider the probability that a game started in a square (i_0, j_0) ends in a certain corner (i^*, j^*). Obviously, $i_0 \leq i^*$ and $j_0 \leq j^*$ for a positive probability. There are many ways to make this trip, and in order to classify these conveniently, consider two fixed lists of rows $i_0, \ldots, i_r$, and of columns $j_0, \ldots, j_s$, where $i_0 < i_1 < \cdots < i_r = i^*$ and $j_0 < j_1 < \cdots < j_s = j^*$ and now contemplate only those moves in which the rook, starting at (i_0, j_0), in each move goes either to the next row or the next column in the respective list; the corner will be reached after exactly $r + s$ moves. There are many ways to do this, and we calculate all at once the total probabilities of the rook following one of this set of paths:

$$(23) \quad \mathscr{P}\left(\{i_0, \ldots, i_{r-1}\}, \{j_0, \ldots, j_{s-1}\}\right) = \prod_{l=0}^{r-1} \frac{1}{h_{i_l j^*} - 1} \prod_{l=0}^{s-1} \frac{1}{h_{i^* j_l} - 1}$$

(empty products interpreted as 1).

In fact, starting from (i_0, j_0) one may move only to (i_1, j_0) or to (i_0, j_1);

the probability for either choice is $1/(h_{i_0 j_0} - 1)$. Now, by induction on n the probability of getting from (i_1, j_0) using only the remaining rows $(i_1, \ldots, i_r)$ and the columns $(j_0, \ldots, j_s)$ is $\mathscr{P}(\{i_1, \ldots, i_{r-1}\}, \{j_0, \ldots, j_{s-1}\})$, as given by (23), because the trip from (i_1, j_0) to (i^*, j^*) may be considered as a walk in a smaller tableau π' obtained from π by removing at least one row. Similarly, the probability of getting from (i_0, j_1) to (i^*, j^*) using rows $(i_0, \ldots, i_r)$ and the remaining columns $(j_1, \ldots, j_s)$ is $\mathscr{P}(\{i_0, \ldots, i_{r-1}\}, \{j_1, \ldots, j_{s-1}\})$. Note that

$$\mathscr{P}(\{i_1, \ldots, i_{r-1}\}, \{j_0, \ldots, j_{s-1}\})$$
$$= (h_{i_0 j^*} - 1)\, \mathscr{P}(\{i_0, \ldots, i_{r-1}\}, \{j_0, \ldots, j_{s-1}\})$$

$$\mathscr{P}(\{i_0, \ldots, i_{r-1}\}, \{j_1, \ldots, j_{s-1}\})$$
$$= (h_{i^* j_0} - 1)\, \mathscr{P}(\{i_0, \ldots, i_{r-1}\}, \{j_0, \ldots, j_{s-1}\})$$

hence the probability of getting from (i_0, j_0) to (i^*, j^*) using only rows $(i_0, \ldots, i_r)$ and columns $(j_0, \ldots, j_s)$ is

$$\left[\frac{1}{h_{i_0 j_0} - 1} ((h_{i_0 j^*} - 1) + (h_{i^* j_0} - 1))\, \mathscr{P}(\{i_0, \ldots, i_{r-1}\}, \{j_0, \ldots, j_{s-1}\}) \right]$$

This proves the result (23), as $h_{i_0 j_0} - 1 = h_{i^* j_0} + h_{i_0 j^*} - 2$. To calculate the probability of reaching (i^*, j^*) from any square along any permissible path we merely have to sum $\mathscr{P}(R, S)$ over all subsets R of $\{1, \ldots, i^* - 1\}$ and over all subsets S of $\{1, \ldots, j^* - 1\}$ and multiply by the probability, $1/n$, of initially landing on any one square. Hence, this probability $\mathrm{Prob}(i^*, j^*)$ is

$$(24) \qquad \mathrm{Prob}(i^*, j^*) = \frac{1}{n} \sum_R \sum_S \mathscr{P}(R, S)$$
$$= \frac{1}{n} \prod_{i=1}^{i^*-1} \left(1 + \frac{1}{h_{ij^*} - 1} \right) \prod_{j=1}^{j^*-1} \left(1 + \frac{1}{h_{i^* j} - 1} \right);$$

which is precisely equation (22).

The game therefore terminates in a given corner with a probability which agrees with that obtained from the counting formula, and therefore we can choose u.a.r.a Young tableau of shape π by playing the game to completion successively with rooks labeled n, $n - 1, \ldots, 1$.

We claimed earlier, however, that the algorithm proved the counting formula (7) also. To establish this, observe that the expression (22) has been shown to represent the probability that our algorithm halts at a specified corner κ of the shape.

Now the sum of these probabilities over all corners is 1, and so

(25)
$$\sum_{\kappa} \frac{1}{n} \prod' \left(\frac{h}{h-1}\right) = 1$$

However, if $N(\pi)$ is the function defined by (7) then (25) asserts that

$$\sum_{\kappa} \frac{N(\pi - \kappa)}{N(\pi)} = 1$$

i.e., that

(26)
$$N(\pi) = \sum_{\kappa} N(\pi - \kappa)$$

Since (26) is also the recurrence satisfied by the number of tableaux of shape π, and since that number and $N(\pi)$ also agree on the tableau which consists of a single square, it follows that the function $N(\pi)$ of (7) is indeed the number of tableaux of shape π, as claimed.

The complete algorithm follows.

ALGORITHM RANYTB

[Enter with a shape π of order n; Exit with a Young tableau of shape π chosen u.a.r. from among all tableaux of shape π.]

(A) $m \leftarrow n;\ \pi' \leftarrow \pi$.

(B) Place the letter m on a square $\mathscr{S}$ of the shape π', the square being chosen u.a.r.

(C) If $\mathscr{S}$ is a corner of π' go to (D); [Move m to a new square] Choose a square $\mathscr{S}'$ u.a.r. from among the $h - 1$ squares in the hook of $\mathscr{S}$, other than $\mathscr{S}$ itself; Move m to $\mathscr{S}'$; Set $\mathscr{S} \leftarrow \mathscr{S}'$ and go to (C).

(D) If $m = 1$, Exit; $m \leftarrow m - 1;\ \pi' \leftarrow \pi' - \mathscr{S}$; to (B) ∎

The FORTRAN program follows the algorithm closely. The instructions which precede instruction 30 store the conjugate partition to the input partition λ in the array y. This array also will hold the output Y-vector, but as the latter is entered from right to left in the array it can never clash with the (shrinking) conjugate partition.

Instruction 30 is step (A) of the algorithm; instructions 40–41 do step (B) by a simple rejection technique; instructions 70–79 move m to a new square (I, J) as in step (C) above; instructions 80–85 modify the partition and its conjugate to reflect the deletion of the chosen corner, and insert one entry in the output Y-vector; the DO 90 loop restores the input partition prior to return.

The process of selecting a square at random on a given chessboard merits some comment. What we do is to select a random square in a *rectangular* board, of size $k \times \lambda(1)$ where k is the number of parts and $\lambda(1)$ is the largest part. This is the smallest rectangular board which holds the given shape.

If the chosen square lies outside the shape we reject the choice, and select a new square. The expected number of selections required before a square is kept is the ratio of areas

$$\bar{S} = k\lambda(1)/n$$

Now the normal order of magnitude of k and of $\lambda(1)$ is $C \sqrt{n} \log n$, consequently we may expect that averaged over all partitions of order n, the statistic $\bar{S}$ will be $O(n^\epsilon)$ for every $\epsilon > 0$, though for some partitions $\bar{S}$ can be quite large.

SUBROUTINE SPECIFICATIONS (NEXYTB)

(1) *Name of subroutine:* NEXYTB.

(2) *Calling statement:* CALL NEXYTB(N,LAMBDA,Y,MTC).

(3) *Purpose of subroutine:* Supplies the sequence of Young tableaux of given shape.

(4) *Description of variables in calling statement:*

Name	Type	I/O/W/B	Description
N	INTEGER	I	The integer which is partitioned.
LAMBDA	INTEGER(N)	I/O	LAMBDA(I) is the Ith part of the partition‡
Y	INTEGER(N)	I/O	Y(I) is the row containing I in the output tableau (I=1,N).
MTC	LOGICAL	I/O	=.FALSE. on input, starts a new sequence; on output, signals end of current sequence.

‡ See program description above for complete description.

(5) *Other routines which are called by this one:* None.

(6) *Approximate number of* FORTRAN *instructions:* 34.

(7) *Remarks:* An extra "part" =0 must be inserted after the last part of the partition LAMBDA, on input, if called with MTC=.FALSE., except that if LAMBDA has N parts, this is not done.

```
      SUBROUTINE NEXYTB(N,LAMBDA,Y,MTC)
      LOGICAL MTC
      INTEGER LAMBDA(N),Y(N),R,S,S1,T
      T=N
      IF(.NOT.MTC) GO TO 40
20    LAMBDA(1)=1
      DO 21  I=2,N
21    LAMBDA(I)=0
      DO 22  J=2,N
      LAMBDA(Y(J))=LAMBDA(Y(J))+1
      IF(Y(J).LT.Y(J-1)) GO TO 30
22    CONTINUE
      MTC=.FALSE.
      RETURN
30    T=LAMBDA(1+Y(J))
      I=N
31    IF(LAMBDA(I).EQ.T) GO TO 32
      I=I-1
      GO TO 31
32    Y(J)=I
      LAMBDA(I)=LAMBDA(I)-1
      T=J-1
40    L=1
43    R=1
42    IF(R.GT.N) GO TO 45
      IF(LAMBDA(R).EQ.0) GO TO 41
      Y(L)=R
      LAMBDA(R)=LAMBDA(R)-1
      L=L+1
      R=R+1
      GO TO 42
41    IF(L.LE.T) GO TO 43
45    IF(N.EQ.1) GO TO 47
      DO 46  J=2,N
      IF(Y(J).LT.Y(J-1)) GO TO 50
46    CONTINUE
47    MTC=.FALSE.
      RETURN
50    MTC=.TRUE.
      RETURN
      END
```

SUBROUTINE SPECIFICATIONS (RANYTB)

(1) *Name of subroutine:* RANYTB.

(2) *Calling statement:* CALL RANYTB(N,LAM,Y).

(3) *Purpose of subroutine:* Selects, u.a.r., a Young tableau of given shape.

(4) *Descriptions of variables in calling statement:*

Name	Type	I/O/W/B	Description
N	INTEGER	I	The integer which is partitioned.
LAM	INTEGER(N)	I	N=LAM(1)+LAM(2)+··· is the input partition of N.
Y	INTEGER(N)	O	Y-vector of the output tableau.

(5) *Other routines which are called by this one:* FUNCTION RAND(I) (random numbers).

(6) *Number of* FORTRAN *instructions:* 31.

```
      SUBROUTINE RANYTB(N,LAM,Y)
      INTEGER LAM(N),Y(N),H
      DO 5  I=1,N
    5 Y(I)=0
      I=0
      L=0
   10 I=I+1
      M=LAM(I)
      DO 20  J=1,M
      Y(J)=Y(J)+1
   20 L=L+1
      IF(L.LT.N) GO TO 10
   30 DO 85 M=1,N
   40 I=1+RAND(1)*Y(1)
      J=1+RAND(1)*LAM(1)
   41 IF(I.GT.Y(J).OR.J.GT.LAM(I)) GO TO 40
   70 H=Y(J)+LAM(I)-I-J
      IF(H.EQ.0) GO TO 80
      L=1+H*RAND(1)
      IF(L.GT.LAM(I)-J) GO TO 60
      J=J+L
      GO TO 70
```

```
60   I=L-LAM(I)+1+J
79   GO TO 70
80   LAM(I)=LAM(I)-1
     Y(J)=Y(J)-1
85   Y(N+1-M)=I
     DO 90   I=1,N
90   LAM(Y(I))=LAM(Y(I))+1
     RETURN
     END
```

SAMPLE OUTPUT

Subroutine NEXYTB was called until MTC=.FALSE. with the input shape

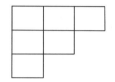

The printed output consisted of the 16 Y-vectors corresponding to the 16 tableaux of this shape, and they are shown below.

```
1  2  3  1  2  1    (40)
1  2  1  3  2  1    (38)
1  1  2  3  2  1    (42)
1  2  1  2  3  1    (44)
1  1  2  2  3  1    (44)
1  2  3  1  1  2    (39)
1  2  1  3  1  2    (47)
1  1  2  3  1  2    (34)
1  2  1  1  3  2    (43)
1  1  2  1  3  2    (46)
1  1  1  2  3  2    (36)
1  2  1  2  1  3    (45)
1  1  2  2  1  3    (42)
1  2  1  1  2  3    (32)
1  1  2  1  2  3    (36)
1  1  1  2  2  3    (32)
```

Subroutine RANYTB was called 640 times with the input shape shown above. For each of the 16 tableaux of this shape, the frequency with which it was obtained is also shown above, next to the Y vector of its shape, in parentheses.

The observed value of χ^2 is 9.0. In 95% of such experiments the value of χ^2 would lie between 6.2 and 27.7 if the tableaux were indeed selected uniformly at random.

Part 2
Combinatorial Structures

15

Sorting (HPSORT/EXHEAP)

A frequently occurring problem in combinatorial work is the *sorting* of an array. One is given $b_1, b_2, \ldots, b_n$, and it is required to permute the members of the array so that the output is in nondecreasing order of size. This is an intensively studied subject, but, even so, important advances continue to be made.

There are various criteria by which one may evaluate the effectiveness of a sorting method, such as (a) the *average* amount of labor (pairwise comparisons or displacements of position) required to sort an array of length n; (b) the *maximum* amount of labor required by *some* sequence of length n; (c) the amount of array storage required; (d) the amount by which the method takes advantage of whatever order is already present in the input list; (e) elegance, compactness, universality, etc.

A comprehensive survey of sorting is given by Knuth [K1, Volume III]. We note here only that the best methods now available require about cn to $cn(\log n)^2$ units of labor on the average, and about $cn(\log n)$ to cn^2 at worst. Furthermore, the best methods will need n to $n + O(\log n)$ storage registers (including the input) and, in the case of merging methods, will speed up operation if considerable order is present on input.

No single sorting method optimizes all departments at once. Our selection here is, we think, a good choice if just one general-purpose sort is to be available for combinatorial applications. It requires an *average* of $cn \log n$ operations, a *maximum* of $cn \log n$ operations, *no* array storage other than the input array, and it is extremely elegant, compact, and universal. Its only unfortunate aspect is that in category (d) above, not only does it fail to take advantage of whatever order is already present, it is actually embarrassingly clumsy when the input list is already sorted! More about this later (see Sample Output, p. 142).

Our choice is the "Heapsort" method of Williams and Floyd, which is of quite recent origin (1964).

First, by a *heap* we mean an array $b_1, b_2, \ldots, b_n$ which has the property that

$$(1) \qquad b_{\lfloor j/2 \rfloor} \geqq b_j \quad (1 \leqq \lfloor j/2 \rfloor < j \leqq n)$$

The importance of this idea rests in the fact that if we imagine the elements $b_1, \ldots, b_n$ as being placed at the successive vertices of a binary tree, as in Fig. 15.1 where $n = 11$, then the sequence is a heap if and only if *every "parent" is at least as large as its two "children."* The reader should study this figure and its relationship to (1) carefully before proceeding. The main properties of the parental relationship which we will use are that

$$(2) \qquad \text{The } parent \text{ of } b_j \text{ is } \quad b_{\lfloor j/2 \rfloor} \qquad (2 \leqq j \leqq n)$$

and

$$(3) \qquad \text{The } children \text{ of } b_j \text{ are} \begin{cases} b_{2j} \text{ and } b_{2j+1}, & \text{if } 2j+1 \leqq n \\ b_{2j} \text{ only,} & \text{if } 2j = n \\ \varnothing, & \text{if } 2j > n \end{cases}$$

The Heapsort algorithm is divided into two phases as follows: First, the input array is transformed into a heap, and, second, the heap is sorted into nondecreasing order.

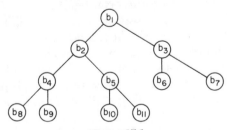

Figure 15.1

The first problem, then, concerns the transformation of a given array into a heap by rearrangement of its elements. The vertices of the tree are processed in reverse order beginning with the first parent, which is $b_{\lfloor n/2 \rfloor}$. Inductively, suppose that we have arrived at a certain parent b_l, and that the left subtree at b_l and the right subtree at b_l have already been transformed into heaps, as shown in Fig. 15.2. How can we make the tree rooted at b_l into a heap?

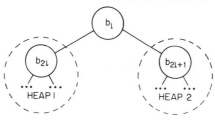

Figure 15.2

We first move b_l to a "safe place," say b^*, thereby creating a vacancy in the tree. Next we begin a "percolating-up" process (Fig. 15.3). The larger of the two descendants of the now-vacant space

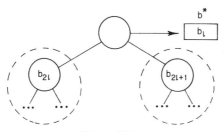

Figure 15.3

moves up if it is larger than b^* and fills the vacancy, but thereby creates another empty slot. Next, the larger of the two descendants of the newly vacated space moves up, if it is larger than b^*, and fills that vacancy but creates another one, and so forth.

The upward motion ceases when the current vacancy has no descendant larger than b^*, and, in particular, it halts if the vacancy has no descendants at all. When it halts, the contents of b^* are moved into the current vacant slot and the upward percolation is complete. At this time the situation in Fig. 15.2 will have been processed, and the full tree shown, including b_l, will constitute a heap. We next go to process b_{l-1} in a similar way.

It is important to break off this piece of the Heapsort as a separate algorithm, which we do as follows:

Definition $\mathscr{F}(l, n)$ is the operation which, given a vertex l in a binary tree of n vertices, and given also that the left subtree at l and the right subtree at l are heaps, carries out an upward percolation process until the entire tree rooted at l is a heap.

ALGORITHM $\mathscr{F}(l, n)$

(A) $l_1 \leftarrow l$; $b^* \leftarrow b_l$.
(B) $m \leftarrow 2l_1$; If $m > n$, to (E); If $m = n$, to (D).
(C)   If $b_{m+1} > b_m$, $m \leftarrow m + 1$.
(D)   If $b^* \geqq b_m$, to (E); $b_{l_1} \leftarrow b_m$; $l_1 \leftarrow m$; to (B).
(E) $b_{l_1} \leftarrow b^*$; Exit ∎

In terms of this Algorithm, the entire transformation of a linear array $b_1, \ldots, b_n$ to a heap is done by

ALGORITHM TOHEAP

(A)   For $l = \lfloor n/2 \rfloor, \lfloor n/2 \rfloor - 1, \ldots, 1$: Do $\mathscr{F}(l, n)$ ∎

We now consider the second phase of Heapsort, in which we sort a heap into nondecreasing order. Here, the reason for dealing with operation $\mathscr{F}(l, n)$ as a separate algorithm will become clear because we will use the same algorithm in a different way in this phase.

First, since $b_1, \ldots, b_n$ now constitute a heap, surely b_1 is the largest of all of the array elements. We therefore exchange b_1 and b_n, at which point the nth array element has its final form. Future operations will therefore leave b_n untouched, and we must now contend with the fact that the reduced tree $b_1, \ldots, b_{n-1}$ no longer is a heap because we just ruined everything by moving the last element to b_1. However, the left subtree at vertex 1 is still a heap, and the right subtree at vertex 1 is still a heap. Hence if we apply operation $\mathscr{F}(1, n - 1)$, then all will be well again!

After doing $\mathscr{F}(1, n - 1)$, the largest of $b_1, \ldots, b_{n-1}$ will now occupy position 1. We exchange b_1 with b_{n-1}, apply $\mathscr{F}(1, n - 2)$, etc., after which the input array appears in sorted order.

The algorithm for sorting a heap is therefore

ALGORITHM SORTHEAP

(A) $n_1 \leftarrow n$.

(B) Exchange b_1, b_{n_1}; If $n_1 \leq 2$, exit; $n_1 \leftarrow n_1 - 1$; Do $\mathscr{F}(1, n_1)$; To (B) ∎

We can observe that Algorithm $\mathscr{F}(l, n)$ is used by both of the phases of Heapsort. It will therefore be written as a subroutine within a subroutine and will be called by the other phases.

The labor involved in $\mathscr{F}(l, n)$ is, at most, one comparison and one displacement for each level of the tree below the level of the lth element. The lth element is at level $1 + \lfloor \log_2 l \rfloor$ in the tree, so $\mathscr{F}(l, n)$ involves, at most,

$$(1 + \lfloor \log_2 n \rfloor) - (1 + \lfloor \log_2 l \rfloor) = O\left(\log \frac{n}{l}\right)$$

comparisons and a like number of displacements of position.

It follows, by summation, that phase Toheap entails at most $O(n)$ operations. Sortheap takes $O(n \log n)$ operations. The full algorithm Heapsort, therefore, is accomplished with at most $O(n \log n)$ comparisons and displacements.

The program is very simply related to the algorithms: Instructions in lines 3–7 are Toheap, 8–10 are Sortheap, and 11–21 do operation $\mathscr{F}(\text{L}, \text{N1})$. The entire program requires just 23 FORTRAN instructions.

To simplify certain applications, a second program, EXHEAP, is given, which performs the same function as HPSORT, but the operations on the data, namely the comparison of two elements, or the transposition of two elements, are performed *outside* the subroutine. This enables the user to operate on more complicated data than just simple numbers. For example, he may wish to put the rows of a matrix into lexicographical order. A slight variation on this idea is found in NETFLO (Chapter 22).

Parameters to EXHEAP are N, INDEX, I, J, ISGN, where N is the length of the list to be sorted, and INDEX is initially set to 0 by the user. If, upon return, INDEX>0, the user is required to interchange items I and J and to reenter the subroutine. (Leave I, J, INDEX unchanged.) If INDEX<0, the user is to compare items I and J, and to enter into ISGN a negative value if item I is to precede item J and a positive value otherwise, then reenter the subroutine, again without disturbing I, J, or INDEX. Finally, if INDEX=0, the sorting is done.

SUBROUTINE SPECIFICATIONS (HPSORT)

(1) *Name of subroutine:* HPSORT.

(2) *Calling statement:* CALL HPSORT(N,B)

(3) *Purpose of subroutine:* Sort a linear array into nondecreasing order.

(4) *Descriptions of variables in calling statement:*

Name	Type	I/O/W/B	Description
N	INTEGER	I	Length of input array.
B	INTEGER(N)	I/O	B(I) is the Ith element of the input array, and then is the Ith element of the sorted array (I=1,N).

(5) *Other routines which are called by this one:* None.

(6) *Number of* FORTRAN *instructions:* 23.

(7) *Remarks:* If B is a REAL array, change type declaration of B,BSTAR to REAL.

```
      SUBROUTINE HPSORT(N,B)
      INTEGER B(N),BSTAR
      N1=N
      L=1+N/2
11    L=L-1
      BSTAR=B(L)
      GO TO 30
25    BSTAR=B(N1)
      B(N1)=B(1)
29    N1=N1-1
30    L1=L
31    M=2*L1
      IF(M-N1) 32,33,37
32    IF(B(M+1).GE.B(M)) M=M+1
33    IF(BSTAR.GE.B(M)) GO TO 37
      B(L1)=B(M)
      L1=M
      GO TO 31
37    B(L1)=BSTAR
      IF(L.GT.1) GO TO 11
      IF(N1.GE.2) GO TO 25
      RETURN
      END
```

SUBROUTINE SPECIFICATIONS (EXHEAP)

(1) *Name of subroutine:* EXHEAP.
(2) *Calling statement:* CALL EXHEAP(N,INDEX,I,J,ISGN).
(3) *Purpose of subroutine:* Sort a list of any items into linear order.
(4) *Descriptions of variables in calling statement:*

Name	Type	I/O/W/B	Description
N	INTEGER	I	Length of input list.
INDEX	INTEGER	I/O	See text for descriptions.
I	INTEGER	O	See text for descriptions.
J	INTEGER	O	See text for descriptions.
ISGN	INTEGER	I	See text for descriptions.

(5) *Other routines which are called by this one:* None.
(6) *Number of* FORTRAN *instructions:* 32.

```
      SUBROUTINE EXHEAP(N,INDEX,I,J,ISGN)
      IF(INDEX) 90,10,80
10    N1=N
      L=1+N/2
20    L=L-1
30    L1=L
40    I=L1+L1
      IF(I-N1) 50,60,70
50    J=I+1
      INDEX=-2
      RETURN
60    J=L1
      L1=I
      INDEX=-1
      RETURN
70    IF(L .GT. 1) GO TO 20
      IF(N1 .EQ. 1) GO TO 110
      I=N1
      N1=N1-1
      J=1
      INDEX=1
      RETURN
80    IF(INDEX-1) 30,30,40
90    IF(INDEX .EQ. -1) GO TO 100
      IF(ISGN .LT. 0) I=I+1
```

```
         GO TO 60
100   IF(ISGN .LE. 0) GO TO 70
         INDEX=2
         RETURN
110   INDEX=0
         RETURN
         END
```

SAMPLE OUTPUT

We illustrate the workings of HPSORT by showing it in the case where, on input, B(I)=I (I=1,10), so that no sorting is really required at all. The first nine lines of output show the status of the array B on input, and then after each displacement. At the ninth step the input array has been transformed into a heap, which concludes the TOHEAP part of the operation. In each line the appearance of a box □ indicates the position of the vacancy in the percolation-up process, and the number within the box is the current content of the "safekeeping" register b^*.

The next 22 lines of output show the transformation of the heap into a fully sorted array. Note how the final output appears from right to left starting from the last array element.

```
 1    2    3    4   |5|   6    7    8    9   10
 1    2    3   |4|  10    6    7    8    9    5
 1    2   |3|   9   10    6    7    8    4    5
 1   |2|   7    9   10    6    3    8    4    5
 1   10    7    9   |2|   6    3    8    4    5
 1   10    7    9    5    6    3    8    4   |2|
10   |1|   7    9    5    6    3    8    4    2
10    9    7   |1|   5    6    3    8    4    2
10    9    7    8    5    6    3   |1|   4    2

 2    9    7    8    5    6    3    1    4   10
 9   |2|   7    8    5    6    3    1    4   10
 9    8    7   |2|   5    6    3    1    4   10
 9    8    7    4    5    6    3    1   |2|  10
 2    8    7    4    5    6    3    1    9   10
 8   |2|   7    4    5    6    3    1    9   10
 8    5    7    4   |2|   6    3    1    9   10
 1    5    7    4    2    6    3    8    9   10
```

7	5	[1]	4	2	6	3	8	9	10
7	5	6	4	2	[1]	3	8	9	10
[3]	5	6	4	2	1	7	8	9	10
6	5	[3]	4	2	1	7	8	9	10
[1]	5	3	4	2	6	7	8	9	10
5	[1]	3	4	2	6	7	8	9	10
5	4	3	[1]	2	6	7	8	9	10
[2]	4	3	1	5	6	7	8	9	10
4	[2]	3	1	5	6	7	8	9	10
[1]	2	3	4	5	6	7	8	9	10
3	2	[1]	4	5	6	7	8	9	10
[1]	2	3	4	5	6	7	8	9	10
2	[1]	3	4	5	6	7	8	9	10
1	2	3	4	5	6	7	8	9	10

16

The Cycle Structure of a
Permutation (CYCLES)

Let σ be a permutation of n letters, and suppose that

$$(1) \qquad\qquad \sigma = t_1 t_2 \cdots t_p$$

where the t_i are transpositions. Then, by the *sign of* σ we mean $+1$ or -1 depending on whether p is even or odd, respectively. It is well known that the sign of σ depends only on σ and not on the particular representation (1), i.e., no matter how we exhibit σ in the form (1), the parity of p is constant.

Computationally, there are better ways of calculating the sign of σ than the above. We have, indeed the following

Theorem Let σ be a permutation of n letters, with q cycles. Then

$$(2) \qquad\qquad \text{sign}(\sigma) = (-1)^{n-q}$$

To prove this, decompose σ into disjoint cyclic permutations

$$(3) \qquad \sigma = C(n_1) C(n_2) \cdots C(n_q), \qquad n_1 + \cdots + n_q = n$$

and observe that each cyclic permutation

$$(4) \qquad\qquad C(m)\colon i_1 \to i_2 \to \cdots \to i_m \to i_1$$

can be written as a product of $m - 1$ transpositions

$$t_{ij}: i \to j \to i$$

as follows

(5) $$C(m) = t_{i_1 i_m} \cdots t_{i_1 i_3} t_{i_1 i_2}$$

Now, substitute (5), with $m = n_1, \ldots, n_q$, in (3). We then see σ written as a product of

$$(n_1 - 1) + \cdots + (n_q - 1) = n - q \text{ transpositions}$$

The calculation of the sign of a permutation is just one application of the cyclic decomposition. Another one arises when an array, say b_i ($i = 1, n$) has to be permuted, so that the new value of $b_{\sigma(i)}$ equals the old value of b_i. It would be impossible to program all forms of this application in one neat, little program, but it is possible to *tag* a permutation to facilitate these applications. Suppose σ is decomposed as in (3), with each $C(m)$ of the form (4). Then among $i_1, \ldots, i_m$ there is a smallest number; we may assume it is i_1. Then $\sigma(i_1) = i_2, \ldots, \sigma(i_m) = i_1$. To tag σ is to replace $\sigma(i_1)$ by $-i_2$. Then, a program which wants to traverse the cycles of σ knows when a cycle is completed, and when it encounters the first element of a new cycle.

One application of such tagging occurs in the *in place* inversion of a permutation. The permutation σ is then replaced by σ^{-1}, where $\sigma^{-1}(i) = j$ iff $\sigma(j) = i$. The execution of this change follows the cycles, replacing $C(m)$ in (4) by

(6) $$C'(m): i_1 \to i_m \to \cdots \to i_2 \to i_1$$

ALGORITHM TAG

[Input n, σ; output: σ with the sign of $\sigma(i)$ reversed for each $\sigma(i)$ for which i is the smallest element in its cycle; n_c, the number of cycles of σ; sgn $= \pm 1$, the sign of σ]

(A) $n_c \leftarrow n$; $i \leftarrow 1$.

(B) $i_1 \leftarrow \sigma(i)$.

(C) If $i_1 \le i$, to (D); $n_c \leftarrow n_c - 1$; $i_2 \leftarrow \sigma(i_1)$; $\sigma(i_1) \leftarrow -i_2$; to (C).

(D) $\sigma(i) \leftarrow -\sigma(i)$; $i \leftarrow i + 1$; if $i \le n$, to (B); sgn $= (-1)^{n-n_c}$; Exit ∎

When following this algorithm on an example, the reader will see how, in each cycle, the signs of all members are reversed in (C); except the first element in each cycle. This serves as a warning that these elements have already been dealt with when a new cycle is sought. Then, in (**D**), all signs are reversed, leaving only the desired tags.

If only n_c and sgn are desired, we replace $\sigma(i) \leftarrow -\sigma(i)$ in (**D**) by $\sigma(i) \leftarrow |\sigma(i)|$.

To invert a permutation, we first tag it; then we reverse each cycle. Let i_0, $i_1 = \sigma(i_0)$, $i_2 = \sigma(i_1)$ be successive members, then setting $\sigma(i_1) \leftarrow i_0$, followed by $i_0 \leftarrow i_1$; $i_1 \leftarrow i_2$, etc., constitutes the core of the process.

ALGORITHM INVERT

[Input a tagged permutation σ; output σ^{-1}, untagged.]

(**A**) $i \leftarrow 0$.
(**B**) $i \leftarrow i + 1$; if $i > n$, Exit; $i_1 \leftarrow -\sigma(i)$; if $i_1 < 0$, to (**B**); $i_0 \leftarrow i$.
(**C**) $i_2 \leftarrow \sigma(i_1)$; $\sigma(i_1) \leftarrow i_0$; if $i_2 < 0$, to (**B**); $i_0 \leftarrow i_1$; $i_1 \leftarrow i_2$; to (**C**) ∎

The FORTRAN program CYCLES performs these tasks, depending on a parameter OPTION, as shown in Table 16.1.

Table 16.1

OPTION	n_c, sgn calculated?	Output	
		Tagged?	Inverted?
−1	yes	no	yes
0	yes	no	no
1	yes	yes	no

SUBROUTINE SPECIFICATIONS

(1) *Name of subroutine:* CYCLES.
(2) *Calling statement:* CALL CYCLES(SIGMA,N,SIGN,NCYCL, OPTION).
(3) *Purpose of subroutine:* Count cycles, find sign of a permutation, tag and/or invert.
(4) *Descriptions of variables in calling statement:*

Name	Type	I/O/W/B	Description
SIGMA	INTEGER(N)	I/O	SIGMA(I) is the value of a permutation $\sigma(I)$ (I=1,N).
N	INTEGER	I	The number of letters being permuted.
SIGN	INTEGER	O	+1 if σ is even, −1 if σ is odd.
NCYCL	INTEGER	O	Number of cycles of the permutation σ.
OPTION	INTEGER	I	See Table 16.1.

(5) *Other routines which are called by this one:*   None.
(6) *Number of* FORTRAN *instructions:*   29.

```
      SUBROUTINE CYCLES(SIGMA,N,SIGN,NCYCL,OPTION)
      INTEGER SIGMA(N),SIGN,OPTION
      IS=1
      NCYCL=N
      DO 5  I=1,N
      I1=SIGMA(I)
6     IF(I1 .LE. I) GO TO 7
      NCYCL=NCYCL-1
      I2=SIGMA(I1)
      SIGMA(I1)=-I2
      I1=I2
      GO TO 6
7     IF(OPTION .NE. 0) IS=-ISIGN(1,SIGMA(I))
5     SIGMA(I)=ISIGN(SIGMA(I),IS)
      SIGN=1-2*MOD(N-NCYCL,2)
      IF(OPTION.GE.0) RETURN
      DO 10 I=1,N
      I1=-SIGMA(I)
      IF(I1 .LT. 0) GO TO 10
      IO=I
15    I2=SIGMA(I1)
      SIGMA(I1)=IO
      IF(I2 .LT. 0) GO TO 10
      IO=I1
      I1=I2
      GO TO 15
10    CONTINUE
      RETURN
      END
```

SAMPLE OUTPUT

If a sequence of edges e_1, e_2, . . . , e_E is an Euler circuit of a graph G, we call the circuit *even* or *odd* depending on whether the permutation σ of E letters

$$\sigma: i \rightarrow e_i \quad (i = 1, E)$$

is an even or odd permutation. For the graph G (Fig. 16.1) we used

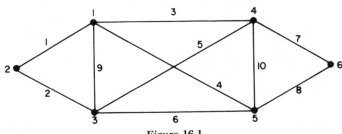

Figure 16.1

the BACKTR subroutine of Chapter 27(C) to generate all Euler circuits, and then used CYCLES to determine if each circuit was even or odd, and the number of cycles in the corresponding permutation.

Below we see, for each of the 44 circuits, the sign of the circuit, the number of cycles in the permutation, and its sequence of edges. Of these, 22 circuits are even, and 22 are odd.

−1	5	1	2	9	4	10	7	8	6	5	3
−1	5	1	2	9	4	10	5	6	8	7	3
1	4	1	2	9	4	8	7	10	6	5	3
−1	5	1	2	9	4	8	7	5	6	10	3
1	6	1	2	9	4	6	5	10	8	7	3
−1	7	1	2	9	4	6	5	7	8	10	3
1	4	1	2	9	3	10	8	7	5	6	4
1	4	1	2	9	3	10	6	5	7	8	4
−1	3	1	2	9	3	7	8	10	5	6	4
1	4	1	2	9	3	7	8	6	5	10	4
−1	5	1	2	9	3	5	6	10	7	8	4
1	6	1	2	9	3	5	6	8	7	10	4
−1	3	1	2	6	10	7	8	4	9	5	3
1	4	1	2	6	10	7	8	4	3	5	9
−1	5	1	2	6	10	5	9	4	8	7	3

−1	5	1	2	6	10	5	9	3	7	8	4
1	4	1	2	6	10	3	9	5	7	8	4
1	4	1	2	6	10	3	4	8	7	5	9
1	4	1	2	6	8	7	10	4	9	5	3
−1	3	1	2	6	8	7	10	4	3	5	9
1	4	1	2	6	8	7	5	9	4	10	3
−1	3	1	2	6	8	7	5	9	3	10	4
1	4	1	2	6	8	7	3	9	5	10	4
1	4	1	2	6	8	7	3	4	10	5	9
−1	5	1	2	6	4	9	5	10	8	7	3
1	6	1	2	6	4	9	5	7	8	10	3
−1	5	1	2	6	4	3	10	8	7	5	9
1	4	1	2	6	4	3	7	8	10	5	9
1	4	1	2	5	10	8	7	3	9	6	4
−1	3	1	2	5	10	8	7	3	4	6	9
1	4	1	2	5	10	6	9	4	8	7	3
1	4	1	2	5	10	6	9	3	7	8	4
−1	5	1	2	5	10	4	9	6	8	7	3
−1	5	1	2	5	10	4	3	7	8	6	9
−1	3	1	2	5	7	8	10	3	9	6	4
1	4	1	2	5	7	8	10	3	4	6	9
1	4	1	2	5	7	8	6	9	4	10	3
−1	5	1	2	5	7	8	6	9	3	10	4
−1	3	1	2	5	7	8	4	9	6	10	3
−1	3	1	2	5	7	8	4	3	10	6	9
1	4	1	2	5	3	9	6	10	7	8	4
−1	5	1	2	5	3	9	6	8	7	10	4
1	6	1	2	5	3	4	10	7	8	6	9
−1	5	1	2	5	3	4	8	7	10	6	9

17

Renumbering Rows and Columns
of an Array (RENUMB)

In this chapter, we study a question which at first sight seems trivial
but which in fact is quite substantial. We are given an $m \times n$ matrix
A and two permutations

$$\sigma: \{1, \ldots, m\} \rightarrow \{1, \ldots, m\}$$
$$\tau: \{1, \ldots, n\} \rightarrow \{1, \ldots, n\}$$

We are asked to renumber the rows and columns of A in accordance
with the given permutations. More precisely, we are to construct the
matrix $\tilde{A}$ whose elements are

(1) $(\tilde{A})_{i,j} = (A)_{\sigma^{-1}(i), \tau^{-1}(j)}$ $(i = 1, \ldots, m; j = 1, \ldots, n)$

Yet there seems to be no problem at all, since the one-step
"algorithm"

(A)   For $i = 1, \ldots, m; j = 1, \ldots, n$: $\tilde{A}(\sigma(i), \tau(j)) \leftarrow A(i, j)$ ∎

does the whole job. Questions of interest arise, however, when we
insist that *no extra array storage* beyond the input data arrays σ, τ, A
be used; in other words, A must be rearranged *in place*. Let us consider a few possible approaches to this problem.

First, we could permute the rows of A according to σ, then permute the columns by τ. This could easily be programmed to use no extra arrays (try it). The disadvantage of this procedure is that each element gets moved twice, an inefficiency.

In the same vein, we could factor the given permutations as products of transpositions, then carry out the successive interchanges of rows or of columns. In this approach each element will be moved several times. For example, let

$$A = \begin{pmatrix} a_{11} & a_{12} & a_{13} \\ a_{21} & a_{22} & a_{23} \\ a_{31} & a_{32} & a_{33} \end{pmatrix}$$

and

$$\sigma = \tau = \{1 \to 3;\ 2 \to 1;\ 3 \to 2\}$$

be the given data. The desired output is the matrix

$$\tilde{A} = \begin{pmatrix} a_{22} & a_{32} & a_{21} \\ a_{32} & a_{33} & a_{31} \\ a_{12} & a_{13} & a_{11} \end{pmatrix}$$

Suppose we observe that

$$\sigma = \tau = t_{12} t_{13}$$

where t_{ij} exchanges letters i and j. Then in this example the number of times each element is moved during the transition from A to $\tilde{A}$ is

$$\begin{pmatrix} 2 & 2 & 3 \\ 2 & 2 & 3 \\ 3 & 3 & 4 \end{pmatrix}$$

We therefore discard this proposal.

A third possibility is this: Store a_{11} in T. Move into a_{11} the element $a_{\sigma^{-1}(1),\, \tau^{-1}(1)}$, then into the latter location move the element which goes there, etc., until the cycle is complete, then move T into the last location in the cycle. Then begin the next cycle, etc. This approach is much closer to what we shall actually do, but it is not deep enough yet, because how, exactly, do we "begin the next cycle?"

What we must do is find some matrix element which was not moved in the first cycle and then follow around the cycle which it belongs to. How do we recognize a matrix element which was not moved in the first cycle?

The matrix elements will have to be flagged in some way when they are moved, and then we can search for an unflagged entry. One possibility is to use a LOGICAL array for just this purpose, but our stipulation regarding no extra array storage would thereby be violated. If it is known in advance that the matrix entries are positive numbers, then we can flag them in the sign position. Aside from restricting the applicability of the subroutine, this approach also doubles the number of times the entries are moved, because the flags must all be reset before exit.

If the matrix elements are restricted to be integers, of either sign, we can (a) double all entries, (b) use the "1's" bit for a flag, (c) halve all entries before exit. The objections to this are as described in the previous paragraph.

We hope now to have convinced the reader that a question of some depth is posed by the requirements of no extra array space, universal applicability of the method, and movement of each matrix entry at most once. We must seek our answer in the direction of understanding the cycle structure of the induced product permutation

$$\rho\colon (i, j) \to (\sigma(i), \tau(j)) \quad (i = 1, \ldots, m; j = 1, \ldots, n)$$

Consider the case where σ and τ are both equal to the permutation $1 \to 4 \to 5 \to 1; 2 \to 3 \to 2$ of five letters. The induced permutation ρ of the product set then has the cycle structure shown below.

$$(1, 1) \to (4, 4) \to (5, 5) \to (1, 1)$$
$$(1, 4) \to (4, 5) \to (5, 1) \to (1, 4)$$
$$(1, 5) \to (4, 1) \to (5, 4) \to (1, 5)$$
$$(2, 2) \to (3, 3) \to (2, 2)$$
$$(2, 3) \to (3, 2) \to (2, 3)$$
$$(1, 2) \to (4, 3) \to (5, 2) \to (1, 3) \to (4, 2) \to (5, 3) \to (1, 2)$$
$$(2, 1) \to (3, 4) \to (2, 5) \to (3, 1) \to (2, 4) \to (3, 5) \to (2, 1)$$

Obviously, within one of its cycles ρ is a cyclic permutation, and to reorder elements under a cyclic permutation is easy. Suppose

$$\alpha_1 \to \alpha_2 \to \alpha_3 \to \cdots \to \alpha_r \to \alpha_1$$

is such a cycle. We can then do

(a) $\quad T \leftarrow \alpha_r$
(b) $\quad \alpha_{i+1} \leftarrow \alpha_i \quad (i = r - 1, r - 2, \ldots, 1)$
(c) $\quad \alpha_1 \leftarrow T$

and the rearrangement is accomplished.

We can, in fact, give a complete description of the cycles of the product permutation ρ in terms of the cycles of σ, τ. If

$$C': \quad i_1 \to i_2 \to \cdots \to i_r \to i_1$$

and

$$C'': \quad j_1 \to j_2 \to \cdots \to j_s \to j_1$$

are, respectively, a cycle of σ and a cycle of τ, let $g = $ g.c.d. (r, s) and let $\lambda = $ l.c.m. (r, s). Corresponding to the pair C', C'' of cycles of σ and τ, there are exactly g different cycles of ρ, each of length λ, namely, the cycles

$$
\begin{array}{ll}
1 & (i_1, j_1) \to (i_2, j_2) \to \cdots \\
2 & (i_1, j_2) \to (i_2, j_3) \to \cdots \\
3 & (i_1, j_3) \to (i_2, j_4) \to \cdots \\
\vdots & \qquad \vdots \qquad\qquad \vdots \\
g & (i_1, j_g) \to (i_2, j_{g+1}) \to \cdots
\end{array}
$$

Furthermore, as C' runs over all cycles of σ and C'' runs through all cycles of τ, we obtain every cycle of ρ as above, each one exactly once.

To use this parameterization of the cycles of ρ in terms of those of σ, τ our first thought might be to proceed as follows (our final thought will differ in a small, but very important, detail):

(a) Run through all elements of one cycle C' of σ, counting them and flagging them, say, by changing the sign of the corresponding entries of the array σ. Let the cycle have r elements.

(b) Do the same as above for one cycle C'' of τ. Let it have s elements.

(c) Calculate $g = $ g.c.d. (r, s); $\lambda = $ l.c.m. (r, s).

(d) For the fixed pair C', C'', move the matrix entries around the set of g cycles of ρ which the pair C', C'' generate, as described above.

(e) By locating the first unflagged (positive) entry of τ, repeat steps (b)–(d) for the next cycle of τ, keeping C' fixed. Continue in this way until all cycles of τ are done (all entries of the array τ are negative).

(f) Reset the entries of τ to positive values. Search the array σ for the first unflagged element, and generate the next cycle C' of σ. Keeping C' fixed, repeat the operations as above.

(g) Proceed until all cycles of σ have been done.

This is, essentially, our method. One difficulty arises if the flagging of elements of σ, τ is done as above. This will become clear if we imagine what happens if the routine is called in the important special case where $\sigma = \tau$, that is, only one linear array is input, and it plays two roles. If we use the sign bits of σ and τ as flags, and if, in fact, σ and τ are the same array, then the logic of the routine will obviously become snarled.

The way out of this problem is to tag, in each cycle of σ, the element $\sigma(i)$ for which i is minimal in that cycle with a minus sign before any actual permuting takes place by using the program CYCLES of Chapter 16. However, once the tags are on, they are not changed during the actual permuting of matrix elements. To find out if $\sigma = \tau$, we first tag σ, then test if $\tau(1)$ is negative; if not, we tag τ also.

The core of the subroutine consists of instructions 50 to 55 in which the cycles of (σ, τ) are traced without s, g, and λ having been explicitly computed; we need only r. Before entering a cycle pair (C', C'') we set

$$k \leftarrow r, \qquad i_1 \leftarrow i, \qquad j_1 \leftarrow j_2 \leftarrow j,$$

where i, j are the tagged elements. Displacements of matrix elements are performed as $i_1 \leftarrow \sigma(i_1)$, $j_1 \leftarrow \tau(j_1)$ until j_1 returns to j_2. At that point, we set $k \leftarrow k - 1$ and test if $i_1 = i$ (return to loop if not) which signifies the completion of a cycle of length λ. We then set $j_1 \leftarrow j_2 \leftarrow \tau(j_1)$ to start a new cycle, provided $k > 0$. Indeed, in each cycle of length λ, k is reduced by λ/s; hence the total number of cycles performed until $k = 0$ is $r/(\lambda/s) = rs/\lambda = g$, and we are finished with the pair (C', C'') of cycles, *without ever having computed g or λ explicitly.*

The number of displacements of matrix elements per cycle pair is, of course, rs; the number of tests is counted as follows. j_1 is tested each time (rs tests); i_1 is tested as k is reduced, (r times). Finally, k is tested g times, so that there is a total of

$$rs + r + g$$

tests. Summed over all pairs (C', C''), this yields

$$mn + m(\text{number of cycles of } \tau) + \sum g$$

In average situations, the last sum is small (though it can be as big as mn) while the average number of cycles of τ is $\log n$ (the maximum is n). In any case, the operation count is $O(mn)$ (see Table 17.1).

Table 17.1 Structure of SUBROUTINE RENUMB

Instruction number	Purpose
10	Outer DO loop through the cycles of σ
20 and two below	The cycle length LC of a cycle of σ is calculated
30	DO loop through the cycles of τ
40	Start of cycle of length λ
50	Continuation of above cycle
55	Test if more cycles required
60, 70	Remove tags from σ, τ

ALGORITHM RENUMB

Given a permutation σ of $1, \ldots, m$; a permutation τ of $1, \ldots, n$; a matrix a_{ij} ($i = 1, \ldots, m$; $j = 1, \ldots, n$); move a_{ij} to $a_{\sigma(i)\tau(j)}$.

(A) Tag σ; If $\tau \neq \sigma$, tag τ; $i \leftarrow 0$.
(B) [*Find next cycle of σ*] $i \leftarrow i + 1$; If $i > m$, to (G); $i_1 \leftarrow -\sigma(i)$; If $i_1 < 0$, to (B); $l \leftarrow 0$.
(C) [*Find length of cycle of σ*] $i_1 \leftarrow \sigma(i_1)$; $l \leftarrow l + 1$; If $i_1 > 0$, to (C); $i_1 \leftarrow i$; $j \leftarrow 0$.
(D) [*Find next cycle of τ*] $j \leftarrow j + 1$; If $j > n$, to (B); If $\tau(j) > 0$, to (D); $j_2 \leftarrow j$; $k \leftarrow l$.
(E) [*Start new product cycle*] $j_1 \leftarrow j_2$; $t_1 \leftarrow a_{i_1 j_1}$.
(F) [*Move matrix elements in one product cycle*] $i_1 \leftarrow |\sigma(i_1)|$; $j_1 \leftarrow |\tau(j_1)|$; $t_2 \leftarrow a_{i_1 j_1}$; $a_{i_1 j_1} \leftarrow t_1$; $t_1 \leftarrow t_2$; If $j_1 \neq j_2$, to (F); $k \leftarrow k - 1$; [*End of product cycle?*] If $i_1 \neq i$, to (F); $j_2 \leftarrow |\tau(j_2)|$; [*All product cycles done?*] If $k \neq 0$, to (E); to (D).
(G) [*Restore arrays*] $\sigma(i) \leftarrow |\sigma(i)|$ ($i = i, \ldots, m$): If $\sigma \neq \tau$, $\tau(j) \leftarrow |\tau(j)|$ ($j = 1, \ldots, n$), Exit ∎

SUBROUTINE SPECIFICATIONS

(1) *Name of subroutine:* RENUMB.
(2) *Calling statement:* CALL RENUMB(M,N,SIG,TAU,A).
(3) *Purpose of subroutine:* Renumber rows and columns of a matrix.
(4) *Descriptions of variables in calling statement:*

Name	Type	I/O/W/B	Description
M	INTEGER	*I*	Number of rows of matrix.
N	INTEGER	*I*	Number of columns of matrix.
SIG	INTEGER(M)	*I*	SIG(I) is the value at I, of the row permutation (I=1,M).
TAU	INTEGER(N)	*I*	TAU(J) is the value, at J, of the column permutation (J=1,N).
A	INTEGER(M,N)	*I/O*	A(I,J) is the I,J entry of the input matrix, and holds Ã(I,J) after execution (I=1,M;J=1,N).

(5) *Other routines which are called by this one:* CYCLES (Chapter 16).

(6) *Number of* FORTRAN *instructions:* 37.

(7) *Remarks:* To use on a noninteger matrix, just change type declaration of A, T1, and T2.

```
      SUBROUTINE RENUMB(M,N,SIG,TAU,A)
      INTEGER SIG(M),TAU(N),A(M,N),T1,T2
      CALL CYCLES(SIGMA,M,IS,NC,1)
      IF(TAU(1).GT.0) CALL CYCLES(TAU,N,IS,NC,1)
      DO 10  I=1,M
      I1=-SIG(I)
      IF(I1.LT.0) GO TO 10
      LC=0
   20 I1=SIG(I1)
      LC=LC+1
      IF(I1.GT.0) GO TO 20
      I1=I
      DO 30  J=1,N
      IF(TAU(J).GT.0) GO TO 30
      J2=J
      K=LC
   40 J1=J2
      T1=A(I1,J1)
   50 I1=IABS(SIG(I1))
      J1=IABS(TAU(J1))
      T2=A(I1,J1)
      A(I1,J1)=T1
      T1=T2
      IF(J1.NE.J2) GO TO 50
      K=K-1
      IF(I1.NE.I) GO TO 50
```

```
      J2=IABS(TAU(J2))
55    IF(K.NE.0) GO TO 40
30    CONTINUE
10    CONTINUE
      DO 60   I=1,M
60    SIG(I)=IABS(SIG(I))
      IF(TAU(1).GT.0) RETURN
      DO 70   J=1,N
70    TAU(J)=IABS(TAU(J))
      RETURN
      END
```

SAMPLE OUTPUT

Subroutine RENUMB was called, with M=N=9, SIG as shown in the first line below, TAU as shown on the second line. The input 9×9 matrix A appears next, followed by the output rearrangement of A.

2	3	9	6	7	8	5	4	1
3	4	5	6	7	8	9	1	2

23	24	25	26	27	28	29	21	22
33	34	35	36	37	38	39	31	32
93	94	95	96	97	98	99	91	92
63	64	65	66	67	68	69	61	62
73	74	75	76	77	78	79	71	72
83	84	85	86	87	88	89	81	82
53	54	55	56	57	58	59	51	52
43	44	45	46	47	48	49	41	42
13	14	15	16	17	18	19	11	12

11	12	13	14	15	16	17	18	19
21	22	23	24	25	26	27	28	29
31	32	33	34	35	36	37	38	39
41	42	43	44	45	46	47	48	49
51	52	53	54	55	56	57	58	59
61	62	63	64	65	66	67	68	69
71	72	73	74	75	76	77	78	79
81	82	83	84	85	86	87	88	89
91	92	93	94	95	96	97	98	99

18

Spanning Forest of a Graph (SPANFO)

One of the most fundamental questions about a graph concerns its connectivity. G is connected if, whenever u and v are distinct vertices of G, there is a path in G which joins u and v. Even if G is not connected, we can define an equivalence relation on its vertex set: u and v are related if there is a path between them. A *connected component* of G is an equivalence class T of vertices of G, under this equivalence relation, together with all of the edges of G which are incident with some vertex of T.

For example, the graph in Fig. 18.1 of 14 vertices and 10 edges has 7 connected components, namely, the 7 pieces of the drawing.

Computationally, we pose the problem as follows. Imagine that we are given the graph G in the form of an array

$$\epsilon(i, j) \qquad (i = 1, 2; j = 1, E),$$

in which $\epsilon(1, j)$, $\epsilon(2, j)$ are, respectively, the two vertices which are end points of edge $j(j = 1, E)$. For the graph of Fig. 18.1, the array might be this:

$\epsilon(1, j)$:	2	4	1	7	5	2	6	2	3	4
$\epsilon(2, j)$:	3	7	9	11	8	5	10	8	8	11

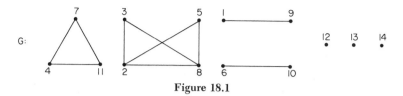

Figure 18.1

We wish to design an algorithm which, given such an array for a graph of n vertices and E edges, will produce the following output:

(a) k, the number of connected components of G,

(b) $x(i)$, the number of the component to which vertex i belongs $(i = 1, n)$,

(c) a rearrangement of the edges in the ϵ array so that listed first are the edges in a spanning tree of component 1, then the edges of a spanning tree of component 2, etc., and finally those edges which are not in any spanning tree,

(d) the sequence of edges in the spanning forest portion of the output list is to have an additional property, namely that each spanning tree can be considered as a directed *rooted* tree, and then its edges are to be listed so that any walk on the tree which begins at the root will follow edges with an ascending sequence of indices. Further, for each edge j, which is in a spanning tree, $\epsilon(1, j)$ is nearer to the root of its tree than $\epsilon(2, j)$, and so, in particular, the root of a tree is $\epsilon(1, r)$, where r is the first edge in that component.

The reader may wish to check the output arrays x, ϵ for the graph of Fig. 18.1, to see how the various properties above are satisfied:

x:	1	2	2	3	2	4	3	2	1	4	3	5	6	7

ϵ:	1	2	8	2	4	4	6	2	3	7
	9	8	5	3	11	7	10	5	8	11

Algorithms for finding such information can be loosely classified into two categories: "on-line" and "off-line." In an on-line algorithm, we think of the computer as being in a room where, from time to time, someone will announce the two endpoints of a certain edge. Without knowing if more edges are to be announced later, the algorithm proceeds to use the latest edge so as to update its picture of the connectivity of the graph. If at any time the procedure is terminated, then with almost no additional computing the desired output items are available. In such a situation the algorithm must deal with the question of maintaining appropriate data structures which can be conveniently updated after a single new edge is announced.

In an off-line algorithm we are presented with a list of *all* of the edges of G before any computation is done. The algorithm is therefore permitted to operate on the entire graph as a unit.

On-line algorithms exist whose operation times are almost, but not quite, linear in the number of edges E of the graph. Off-line algorithms exist whose times are linear in E. Such methods can be formulated in terms of breadth-first-search or in terms of depth-first-search of a graph. In this chapter we describe two algorithms, a depth-first-search algorithm which is due to R. Tarjan, and then another, breadth first, algorithm, which offers economies of array space. A FORTRAN program for the latter algorithm is given.

(B) DEPTH-FIRST-SEARCH

Given G, of n vertices and E edges, we initiate the search by choosing a vertex v of G. We *scan* v by examining each of its neighbors. Any neighbor w of v which has not already been assigned to a connected component of G is assigned to the current component, and the vertex w is placed on top of a stack. Neighbors which have already been assigned are ignored.

After scanning v, we remove the vertex which is then on top of the stack, and call it v. We then scan v, as before. When the stack is empty, we have finished a connected component. We then increment the component number and choose, for v, any vertex which has not yet received a component number. If no such v exists, the algorithm halts.

The reader is invited to try the algorithm by hand on the first two components of Fig. 18.1, for instance.

The algorithm as so far described meets three of the four requirements (a)–(d) which we posed. It finds k, assigns each vertex to a component, and if $v_1, v_2, \ldots, v_l$ is the sequence in which the vertices of a certain component were scanned, then the edge sequence $(v_1, v_2), (v_2, v_3), \ldots, (v_{l-1}, v_l)$ forms a spanning tree for the component which satisfies condition (d).

The ingredient which is so far missing is this: when we are scanning vertex v, and we examine a vertex w in its neighborhood, consider the case where w has already been assigned a component number. The edge (v, w) might be an edge which already is part of the spanning tree, it might be an edge which is not in the tree and which we are seeing for the first time, or it might be such an "extra" edge which we are seeing for the second time.

We can tell if (v, w) is already in the spanning tree simply by remembering the immediate predecessor v' of v in the spanning tree. If $w = v'$, the edge is already in the spanning tree and we can ignore it.

Otherwise, if (v, w) is an "extra" edge we want to store it in the list of extra edges at the right side of the ϵ array if we are now seeing (v, w) for the first time, and to ignore it if we have already seen (v, w).

In order to make this decision some extra information is needed. We assign to each vertex v, not only its component number, but also its position in the scan sequence. Call this $NUM(v)$. Thus, if w was scanned before v was, we shall have $NUM(w) < NUM(v)$ and vice-versa. With the NUM array our decision is easy to make: if $NUM(v) < NUM(w)$ we adjoin (v, w) to the list of extra edges, else we ignore it because it will then be chosen exactly once.

Altogether, then, we ignore (v, w) if

$$\{w = \text{Predecessor}(v) \quad \text{or} \quad NUM(v) > NUM(w)\}$$

and otherwise adjoin it to the list of extra edges.

The formal algorithm follows.

ALGORITHM DEPTHFIRST

(A)   Set $x(v) \leftarrow 0$ (all v); $k \leftarrow 0$; $v \leftarrow 1$; num $\leftarrow 0$.

(B)   [*Scan next v*] If $v > n$, Exit; If $x(v) \neq 0$, set $v \leftarrow v + 1$ and go to (B); [Begin kth component at v] $k \leftarrow k + 1$; num $\leftarrow$ num $+ 1$; $x(v) \leftarrow (k, \text{num})$; Write $(v, 0)$ on stack.

(C)   [*Component k finished?*] If stack is empty, set $v \leftarrow v + 1$ and go to (B); Read $(j, \text{Prev}) \leftarrow$ top of stack; For each vertex $r \in Nbhd(j)$ do:

   (C1)   If $x(r) \neq 0$ then go to (C2); [Vertex r encountered for first time] num $\leftarrow$ num $+ 1$; $x(r) \leftarrow (k, \text{num})$; Top of stack $\leftarrow (r, j)$; List of tree edges $\leftarrow (j, r)$; Next r.

   (C2)   If $r =$ Prev or $x(r) > x(j)$ then do next r; List of extra edges $\leftarrow (j, r)$; Next r.

   End
   Go to (C). ■

(C)   A BREADTH-FIRST ALGORITHM

We turn now to a modified approach, which has the advantage over the depth-first approach that it can be entirely executed within two arrays, the input array $\epsilon(l, m)$ ($l = 1, 2$; $m = 1, E$), which is the edge list of the graph, and the array $x(i)$ ($i = 1, n$) which on output has in $x(i)$ the component to which vertex i belongs.

As before, we scan and visit, but the scanning of vertices takes place in the same order as the visiting. Thus, we are not dealing with a stack but with a so-called *queue*. It so happens that, in this approach, storage can be so used that ϵ and x suffice. Of course, the entries of ϵ and x are used a few times each, for somewhat different purposes.

Since no auxiliary storage is available, the lists of neighbors of each vertex have to be stored in ϵ itself in such a manner that essentially no searching will be needed. We do this by replacing all occurrences of vertex p by a linked list, which points from one such occurrence to another. $x(p)$ points to the first such occurrence, while at the last such occurrence p itself is inserted. The addresses (l, m) in ϵ are encoded as $lM + m$, where M is a sufficiently large number, say $M = 1 + \max(n, E)$. All these quantities are given negative signs to indicate that no scanning or visiting has yet taken place.

ALGORITHM LINK (x, ϵ, n, E)

For $i = 1, n$: $x(i) \leftarrow -i$
For $m = 1, E$:
  For $l = 1, 2$:
    $p \leftarrow \epsilon(l, m)$; $\epsilon(l, m) \leftarrow x(p)$; $x(p) \leftarrow -(lM + m)$ ∎

As an example, consider the graph of Fig. 18.2

Figure 18.2

On input this graph might have been described by the arrays

ϵ

2	4	4	2	1
3	2	3	1	3

x

−	−	−	−

After conversion to linked lists in which we have chosen $M = 10$ for convenience, the arrays ϵ and x look like this:

$$\epsilon \quad \begin{array}{|c|c|c|c|c|} \hline -22 & -13 & -4 & -2 & -1 \\ \hline -23 & -14 & -25 & -15 & -3 \\ \hline \end{array} \qquad x \quad \begin{array}{|c|c|c|c|} \hline -24 & -11 & -21 & -12 \\ \hline \end{array}$$

For example, the linked list of vertex 3 is found by starting at $x(3)$ and walking through ϵ thus:

$$x(3) = (2, 1) \to (2, 3) \to (2, 5) \to 3$$

When the scanning of vertex p has reached $\epsilon(l, m)$, say, we set $s \leftarrow |\epsilon(l, m)|$ and reinsert $\epsilon(l, m) \leftarrow p$ to restore the endpoint list. We then do two things:

1. Visit the neighbor of p at the other endpoint of edge m; that is, we go to $\epsilon(3 - l, m)$ and follow the linked list to the end to find the name q of the neighbor. Note that if q has already been visited no second visit is needed. Note also that if q has already been scanned, qp would be a suitable edge with which to join p to an already partially constructed spanning tree.

2. Examine s. If $s > M$, then s points to another occurrence of p in the edge list, and it will be the next place from which to visit. If $s < M$, then we have just completed the scanning of p. In particular, we shall see to it that if $s = 0$, a component has just been completed. When the scanning of p is complete, we shall signal this fact by replacing $x(p)$ by a positive quantity.

Before we discuss the scanning process any further, we first examine the visiting process in detail.

a) We identify as-yet-unvisited neighbors q of a vertex p that we are scanning. We are also interested in the endpoint (l_1, m_1) of the linked list of q, and we insert 0 into this location. Usually, this zero will be overwritten later, but when it isn't, it signifies that q was the last vertex in the component that was scanned. While we travel along the linked list of q, we change the signs of all pointers, to indicate to all future attempts to visit q that such a visit has already taken place.

b) If the neighbor was already visited, there is no reason to visit it again. Continuing the attempt to visit would lead to useless or even confusing duplicate information, and would adversely affect the labor estimate. So, suppose we are scanning p and have arrived at $\epsilon(l, m)$. We then "cross over", setting $s_1 \leftarrow \epsilon(3 - l, m)$. The following cases can occur:

b_1) $s_1 < -M$. Then s_1 points to the next location of a linked list of the neighbor q. We reinsert $|s_1|$ into ϵ and move on.

b_2) $-M < s_1 < 0$. We have just found $q = |s_1|$; let the location be (l_1, m_1). We set $\epsilon(l_1, m_1) \leftarrow 0$ and remember (l_1, m_1).

b_3) $s_1 = 0$. The vertex has been visited: Abandon visit.

b_4) $0 < s_1 < M$. Then s_1 is the name of a vertex. If $x(s_1) > 0$ then s_1 has been scanned; remember m_1 so we can use this edge for the spanning tree. In any case, abandon the visit.

b_5) $s_1 > M$. Pointer in a linked list, but vertex has already been visited. Abandon.

ALGORITHM VISIT $(x, \epsilon, n, E, q, l_1, m_1, \alpha)$

[Visits, with possibility of abandonment, a vertex starting at $\epsilon(l_1, m_1)$. In each of the four cases b_2, b_3, b_4, b_5 described above, output is as follows:

b_2: Output q (new unvisited neighbor); (l_1, m_1) (the end of the linked list of q); $\alpha = 1$; Insert 0 into last place $\epsilon(l_1, m_1)$ of linked list.

b_3, b_4 (if $x(s_1) \leqq 0$), or b_5: Output $\alpha = 0$.

b_4 (if $x(s_1) > 0$): Output m_1 (edge to scanned neighbor); $\alpha = -1$.

In all cases, change all negative pointers to positive as they are encountered.]

(A) $q \leftarrow \epsilon(l_1, m_1)$;
  If $q < -M$, set $\epsilon(l_1, m_1) \leftarrow -q$; $l_1 \leftarrow \lfloor -q/M \rfloor$; $m_1 \leftarrow -q - l_1 M$;
  To (A).
  If $-M < q < 0$, set $q \leftarrow -q$; $\epsilon(l_1, m_1) \leftarrow 0$; $\alpha \leftarrow 1$; Exit.
  If $0 < q < M$, if $x(q) > 0$, set $\alpha \leftarrow -1$; Exit;
        else set $\alpha \leftarrow 0$; Exit.
  If $q = 0$ or $q > M$, set $\alpha \leftarrow 0$; Exit ■

Now that we have a complete description of a visit, we resume discussion of the scanning of a vertex p. We start traversing the linked list of p at the location encoded in $x(p)$. If p is a root, we first visit p to find the last point (l_0, m_0) of its linked list and to reset signs so that p will never be visited successfully. Only then does the actual scanning start. Let q be the first neighbor of p that is visited successfully; then we want to scan it as soon as we are finished with p. To this effect, we set $\epsilon(l_0, m_0) \leftarrow q$, and save in (l_0, m_0) the end of the linked list of q, which we have just found! We continue like this, hooking up one linked list after the other. The last entry in this list will be 0 (see case b_2), which signifies that the component has been completely scanned.

Meanwhile, we should not forget the main objective, the construction of a spanning tree. All we need for that, while scanning p, is one single edge (say, r) joining p to an already scanned vertex. To effect this we set $r \leftarrow 0$ initially and when we find an edge m_1 (see first case b_4) we set $r \leftarrow m_1$. This chooses just one edge for p, and if by chance $r = 0$ when p has been scanned, then p is a root.

ALGORITHM SCAN (x, ϵ, n, E, p, l_0, m_0, m, r)

[Scans vertex p. If p is a root, r is returned unchanged. Otherwise, edge r joins p to a previously scanned vertex. m is the name (if $m \neq 0$) of the vertex which is listed at the end of the linked list of p.]

(A) $s \leftarrow -x(p)$.
(B) $l \leftarrow \lfloor s/M \rfloor$; $m \leftarrow s - lM$; If $l = 0$, Exit;
  $l_1 \leftarrow 3 - l$; $m_1 \leftarrow m$; VISIT (x, ϵ, n, E, q, l_1, m_1, α)
  If $\alpha = -1$, then $r \leftarrow m_1$;
  If $\alpha = 1$, then $\epsilon(l_0, m_0) \leftarrow q$ and $(l_0, m_0) \leftarrow (l_1, m_1)$;
  $s \leftarrow |\epsilon(l, m)|$; $\epsilon(l, m) \leftarrow p$; to (B) ■

To process a component, we now combine VISIT and SCAN in the proper manner, as described above. In order to assure that edges of the spanning tree end up in the right places, we have a location assigner L, initially zero, which is incremented each time a new edge is found. There is also k, which counts components. When vertex q has been scanned, if it is a root, we set $x(q) \leftarrow k$; if it is not, we register the edge r by entering the other endpoint of it in $x(q)$, q into $\epsilon(2, r)$, and $-L$ into $\epsilon(1, r)$.

ALGORITHM COMPONENT (x, ϵ, n, E, p, k, L)

[Controls the scanning of one component, assigning component number and labeling edges of the spanning tree with future addresses.]

(A) $s \leftarrow -x(p)$; $l_0 \leftarrow \lfloor s/M \rfloor$; $m_0 \leftarrow s - l_0 M$; $r \leftarrow 0$.
  VISIT (x, ϵ, n, E, q, l_0, m_0, α)

(B) SCAN $(x, \epsilon, n, E, q, l_0, m_0, m, r)$
If $r = 0$, then $k \leftarrow k + 1$ and $x(q) \leftarrow k$; else $L \leftarrow L + 1$;
$x(q) \leftarrow \epsilon(1, r) + \epsilon(2, r) - q$; $\epsilon(1, r) \leftarrow -L$; $\epsilon(2, r) \leftarrow q$.
If $m = 0$, Exit; Else $q \leftarrow m$; To **(B)** ∎

To complete the whole construction of a spanning forest, we therefore have to call LINK, deal with isolated vertices separately $(x(i) = -i)$, call COMPONENT for each other unprocessed vertex and then sort the edges and insert component numbers into x. All this has now become quite simple.

ALGORITHM SPANFO (x, ϵ, n, E, k)

[Constructs a spanning forest for a graph according to requirements a–d of the Introduction.]

(A) $M \leftarrow 1 + \max(n, E)$; LINK$(x, \epsilon, n, E)$; $k \leftarrow 0$; $L \leftarrow 0$.
(B) For $i = 1, n$:
  If $x(i) = -i$, then $k \leftarrow k + 1$; $x(i) \leftarrow k$; Next i;
  If $x(i) > 0$, then next i;
  Else COMPONENT$(x, \epsilon, n, E, i, k, L)$; next i.
(C) For $m = 1, E$: perform **(D)**.
(D) If $\epsilon(1, m) > 0$, next m; Else, $r \leftarrow -\epsilon(1, m)$; Interchange
  $\epsilon(2, m) \leftrightarrow \epsilon(2, r)$; $\epsilon(1, m) \leftarrow \epsilon(1, r)$; $\epsilon(1, r) \leftarrow x(\epsilon(2, r))$; to **(D)**.
(E) For $i = 1, L$: $x(\epsilon(2, i)) \leftarrow x(\epsilon(1, i))$ ∎

The FORTRAN implementation packages all procedures into one program with identification of the parts. The algorithms have been followed rather closely though a few shortcuts were made to avoid duplicate codes. Also, the use of α has been replaced by direct transfers (42, 43, 44). The two returns from VISIT, to COMPONENT and to SCAN have been made transparent by the assigned GO TO RTNVIS.

Comment on Array Storage We have observed that one of the principal motivations for our choice of the second algorithm as our FOR-TRAN program was the fact that it utilized only two named arrays: x, ϵ. The alert reader will have noticed that some of these words are "packed" with two pieces of information, and so the question can be raised as to what the "real" array storage requirement is.

To answer this, we count the *bits* of storage, rather than the *words*. The ϵ array consists of $2E$ words, each of which might contain a number as large as $3M$ ($M = 1 + \max(n, E)$) in absolute value, and the sign bit is also used. Hence $2E \log_2 3M$ bits of array storage are used for ϵ. Similarly, the x-array uses about $n \log_2 3M$ bits, and so our array storage requirements total about $(2E + n)\log_2 3M$ bits. For instance, for a "dense" graph with $E = \alpha n^2$ we need $4\alpha n^2 \log_2 n + O(n^2)$ bits, whereas for a "sparse" graph, where $E = \alpha n$, we would use just $(2\alpha + 1)n \log_2 n + O(n)$ bits of array storage. Such bit counts form a valid basis of comparison between the array storage requirements of various proposed methods of solving the same problem.

SUBROUTINE SPECIFICATIONS

(1) *Name of subroutine:*   SPANFO.
(2) *Calling statement:*   CALL SPANFO(N,E,ENDPT,K,X)
(3) *Purpose of subroutine:*   Determine connectivity of a graph; find spanning forest.
(4) *Descriptions of variables in calling statement:*

Name	Type	I/O/W/B	Description
N	INTEGER	I	Number of vertices in input graph G.
E	INTEGER	I	Number of edges in G.
ENDPT‡	INTEGER(2,E)	I/O	ENDPT(1,I),ENDPT(2,I) are the two vertices in edge I(I=1,E).
K	INTEGER	O	Number of connected components in G.
X	INTEGER(N)	O	X(I) is the component to which vertex I belongs (I=1,N).

‡ See text for arrangement of output array ENDPT.

(5) *Other routines which are called by this one:*   None
(6) *Number of* FORTRAN *instructions:*   79.

```
      SUBROUTINE SPANFO(N,E,ENDPT,K,X)
      IMPLICIT INTEGER(A-Z)
      DIMENSION ENDPT(2,E),X(N)
      MM=1+MAXO(N,E)
C
C     START LINK
C
      DO 11  I=1,N
11    X(I)=-I
      DO 12  M=1,E
      DO 12  L=1,2
```

```
      P=ENDPT(L,M)
      ENDPT(L,M)=X(P)
12    X(P)=-L*MM-M
C
C     END LINK--START MAIN S/R
C
      K=0
      LOC=0
      I=0
20    I=I+1
      IF(I .GT. N)GO TO 21
        Q=X(I)
      IF(Q .GT. 0) GO TO 20
      K=K+1
      X(I)=K
      IF(Q+I)30,20,20
C        30 IS CALL TO COMPONENT-FROM WHERE TO 20
21    DO 23  M=1,E
22    R=-ENDPT(1,M)
      IF(R .LT. 0) GO TO 23
      S=ENDPT(2,M)
      ENDPT(2,M)=ENDPT(2,R)
      ENDPT(2,R)=S
      ENDPT(1,M)=ENDPT(1,R)
      ENDPT(1,R)=X(ENDPT(2,R))
      GO TO 22
23    CONTINUE
      DO 24  I=1,LOC
24    X(ENDPT(2,I))=X(ENDPT(1,I))
      RETURN
C
C     END MAIN S/R--START COMPONENT
C
30    P=I
      S=-Q
      ASSIGN 31 TO RTNVIS
      GO TO 53
C     CALL TO VISIT-FROM WHERE TO 31
31    LO=L1
      MO=M1
      R=0
      ASSIGN 43 TO RTNVIS
      GO TO 41
```

```
32   GO TO 40
C    CALL TO SCAN-FROM WHERE TO 33
33   IF(R .EQ. 0) GO TO 34
     LOC=LOC+1
     X(P)=ENDPT(1,R)+ENDPT(2,R)-P
     ENDPT(1,R)=-LOC
     ENDPT(2,R)=P
34   P=M
     IF(M)20,20,32
C
C    END COMPONENT-START SCAN
C
40   S=-X(P)
41   L=S/MM
     M=S-L*MM
     IF(L .EQ. 0) GO TO 33
     L1=3-L
     M1=M
     GO TO 50
C    CALL TO VISIT-FROM WHERE TO 42,43 OR 44
42   R=M1
     GO TO 44
43   ENDPT(LO,MO)=Q
     LO=L1
     MO=M1
44   S=IABS(ENDPT(L,M))
     ENDPT(L,M)=P
     GO TO 41
C
C    END SCAN-START VISIT
C
50   Q=ENDPT(L1,M1)
     IF(Q)51,44,54
51   IF(Q .LT. -MM) GO TO 52
     Q=-Q
     ENDPT(L1,M1)=0
     GO TO RTNVIS,(31,43)
52   ENDPT(L1,M1)=-Q
53   L1= -Q/MM
     M1=-Q-L1*MM
     GO TO 50
54   IF(Q .GT. MM) GO TO 44
     IF(X(Q))44,42,42
```

```
C
C      END VISIT
C
       END
```

SAMPLE OUTPUT

SPANFO was called for the graph of Fig. 18.1. The input and output array ENDPT is shown below together with the output array X and the number of connected components K=7.

The first edge in the output ENDPT array is a spanning tree for the first component (Fig. 18.3). The next three edges are a spanning tree for the second component (Fig. 18.4). The next two span the third

Figure 18.3   Figure 18.4

component (Fig. 18.5). Finally the last one spans the fourth component (Fig. 18.6).

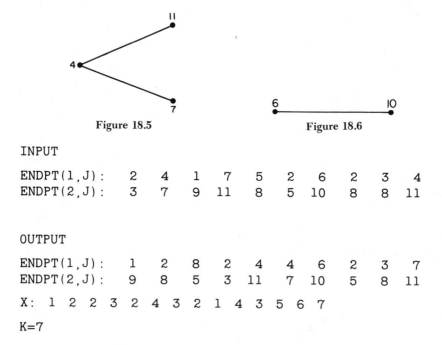

Figure 18.5   Figure 18.6

INPUT

ENDPT(1,J):	2	4	1	7	5	2	6	2	3	4
ENDPT(2,J):	3	7	9	11	8	5	10	8	8	11

OUTPUT

ENDPT(1,J):	1	2	8	2	4	4	6	2	3	7
ENDPT(2,J):	9	8	5	3	11	7	10	5	8	11

X: 1 2 2 3 2 4 3 2 1 4 3 5 6 7

K=7

19

Newton Forms of a Polynomial (POLY)

A polynomial $f(x)$ of degree $n - 1$ is said to be in Newton form with respect to $x_1, \ldots, x_{n-1}$ if it is written as

$$(1) \qquad f(x) = a_1 + a_2(x - x_1) + a_3(x - x_1)(x - x_2) \\ + \cdots + a_n(x - x_1) \cdots (x - x_{n-1})$$

In this chapter, we consider (primarily for use in Chapter 20) algorithms for transforming the coefficients of f from one Newton form to another and related questions.

First, values of f can easily be calculated by rewriting (1) as

$$(2) \qquad f(x) = a_1 + (x - x_1)(a_2 + (x - x_2)(a_3 + (x - x_3) \\ \times (\cdots + (a_{n-1} + (x - x_{n-1})a_n) \cdots)$$

More explicitly, $v = f(z)$ is the result of

ALGORITHM VALUE

[Input: $z, a_1, \ldots, a_n, x_1, \ldots, x_{n-1}$; output $v = f(z)$, where $f(x)$ is given by (1).]

(A) $v \leftarrow a_n.$

(B) $v \leftarrow (z - x_k)v + a_k$ $(k = n - 1, n - 2, \ldots, 1)$ ∎

The following special cases are of interest here:

(i) $x_1 = \cdots = x_{n-1} = 0.$ Then (1) is the usual form of a polynomial as a series of powers of x.

(ii) $x_1 = \cdots = x_{n-1} = c.$ Then (1) is the Taylor expansion of $f(x)$ at c.

(iii) $x_i = i - 1$ $(i = 1, \ldots, n - 1)$. Here $f(x)$ is expanded as

(3) $$f(x) = a_1(x)_0 + a_2(x)_1 + \cdots + a_n(x)_{n-1}$$

in a series in the factorial polynomials

(4) $$(x)_m = x(x - 1) \cdots (x - m + 1); \quad (x)_0 = 1$$

Transitions between the different forms can be obtained by a well-known theorem of interpolation theory (see, for example, Conte and de Boor [CB1, p. 199]) which asserts that the coefficients $b_1, \ldots, b_n$ of the Newton form of f with respect to $z, x_1, \ldots, x_{n-2}$ are found from the coefficients $a_1, \ldots, a_n$ of the Newton form of f with respect to $x_1, \ldots, x_{n-1}$ by means of a simple modification of algorithm VALUE, which stores the intermediate values of v.

ALGORITHM NEWTON

[Input $a_1, \ldots, a_n, x_1, \ldots, x_{n-1}, z$ with $f(x)$ given in its Newton form (1), with respect to $x_1, \ldots, x_{n-1}$, output is $b_1, \ldots, b_n$ the coefficients of $f(x)$ in its Newton form with respect to $z, x_1, \ldots, x_{n-2}.$]

(A) $b_n \leftarrow a_n.$

(B) $b_i \leftarrow a_i + (z - x_i)b_{i+1}$ $(i = n - 1, \ldots, 1)$ ∎

To prove this, rewrite **(A)** and **(B)** as

$$a_n = b_n$$
$$a_i = b_i - (z - x_i)b_{i+1} \quad (i = n - 1, \ldots, 1)$$

and substitute in (1). There are two terms which involve b_i, namely,

$$b_i(x - x_1) \cdots (x - x_{i-1}) - (z - x_{i-1})b_i(x - x_1) \cdots (x - x_{i-2})$$
$$= b_i(x - x_1) \cdots (x - x_{i-2})(x - x_{i-1} - z + x_{i-1})$$
$$= b_i(x - z)(x - x_1) \cdots (x - x_{i-2})$$

which is exactly the term with b_i in the Newton form with respect to $z, x_1, \ldots, x_{n-2}$, as claimed.

As an application, in order to obtain from the polynomial form (i) of $f(x)$ the Taylor expansion (ii) at c, we call NEWTON $n-1$ times, with $z; x_1, \ldots, x_{n-1}$ having the values, respectively,

$$c; \quad 0, 0, \ldots, 0, 0$$
$$c; \quad c, 0, \ldots, 0, 0$$
$$\cdot \quad \cdot$$
$$\cdot \quad \cdot$$
$$\cdot \quad \cdot$$
$$c; \quad c, c, \ldots, c, 0$$

If we take for a_i and b_i the same arrays, then we may skip step (B) of NEWTON every time when $z = x_i$.

ALGORITHM TAYLOR

[Input $c; a_1, \ldots, a_n$, the coefficients of $f(x)$ in power sum form (i); output $a_1, \ldots, a_n$, the coefficients of $f(x)$ in Taylor expansion form (ii).]

(A) For $m = 1, \ldots, n-1$ do:
　　　For $i = n-1, \ldots, m$ do:

$$a_i \leftarrow a_i + c a_{i+1} \quad \blacksquare$$

Observe that if $a_1, \ldots, a_n$ are the coefficients of f in an expansion in powers of $x - d$, then the output will be the coefficients of an expansion in powers of $x - d - c$.

As a second application, we convert $f(x)$ from polynomial form (i) to factorial form (3). Thus, let

(5) $$f(x) = a_1 + a_2 x + \cdots + a_n x^{n-1}$$

then a_1 will not be changed, as

$$f(x) = a_1 + x(a_2 + a_3 x + \cdots + a_n x^{n-2})$$

and our next operation consists of writing

(6) $$a_2 + a_3 x + \cdots + a_n x^{n-2}$$

in the form

(7) $$a_2' + (x - 1)(a_3' + a_4' x + \cdots + a_n' x^{n-3})$$

This requires a call to NEWTON, with $z = 1$, $x_i = 0$, applied to the polynomial (6). Similarly, in the next phase, we apply NEWTON to the paranthesized polynomial in (7), with $z = 2$, $x_i = 0$ to obtain

$$a_3'' + (x - 2)(a_4'' + \cdots + a_n'' x^{n-4})$$

and so on.

ALGORITHM STIRLING

[Input: $a_1, \ldots, a_n$, coefficients of $f(x)$ as power sum (5); output $a_1, \ldots, a_n$, coefficients of $f(x)$ is factorial form (3).]

(A) For $m = 1, \ldots, n - 1$ do:
  $z \leftarrow m - 1$; for $i = n - 1, \ldots, m$ do:

$$a_i \leftarrow a_i + z a_{i+1} \quad \blacksquare$$

The same result would have been obtained by applying Algorithm NEWTON more straightforwardly to all of $a_1, \ldots, a_n$, starting with $x_1 = \cdots = x_{n-1} = 0$, $z = n - 1$; then $x_1 = n - 1$, $x_2 = \cdots = x_{n-1} = 0$, $z = n - 2$; then $x_1 = n - 2$, $x_2 = n - 1$, $x_3 = \cdots = x_{n-1} = 0$, $z = n - 3$, etc. On balance, Algorithm STIRLING above seems more efficient.

If Algorithm STIRLING is applied to $f(x) = x^{n-1}$, the output values of $a_1, \ldots, a_n$ express x^{n-1} as a linear combination of $(x)_0, \ldots, (x)_{n-1}$. These are the *Stirling numbers of the second kind*.

A third application of Algorithm NEWTON will convert factorial series to ordinary power series. If $f(x)$ in the form (3) is given, then we simply call NEWTON $n - 1$ times, with $z; x_1, \ldots, x_{n-1}$ having the following values

$$0; \quad 0, 1, 2, \ldots, n - 2$$
$$0; \quad 0, 0, 1, \ldots, n - 3$$
$$\cdot \quad \cdot$$
$$\cdot \quad \cdot$$
$$\cdot \quad \cdot$$
$$0; \quad 0, 0, 0, \ldots, 1$$

Of course, we may omit again steps (B) in NEWTON for which $z = x_i$, and obtain

ALGORITHM REVERSE STIRLING

[Input $a_1, \ldots, a_n$, coefficients of $f(x)$ in factorial form (3); output $a_1, \ldots, a_n$, coefficients of $f(x)$ in power sum form (5).]

(A)   For $m = 1, \ldots, n - 1$ do:

$z \leftarrow m - n$

For $i = n - 1, \ldots, m$ do:

$$z \leftarrow z + 1; \quad a_i \leftarrow a_i + za_{i+1} \quad \blacksquare$$

If Algorithm REVERSE STIRLING is applied to $f(x) = (x)_{n-1}$ the output values $a_1, \ldots, a_n$ give $(x)_{n-1}$ as a linear combination of $x^0, \ldots, x^{n-1}$. These are the *Stirling numbers of the first kind* (with alternating signs).

Now we want to produce a program which incorporates the algorithm VALUE for cases (i) and (iii), TAYLOR (including the ability to produce just a few of the Taylor coefficients), STIRLING and REVERSE STIRLING. In view of the strong similarities, we want to combine them as well as we can. The following is sufficiently general to accomodate the inner loop of each. It contains a parameter ϵ, which is 0 or 1; and γ, which is True if and only if the intermediate values of v (as in VALUE) are to be saved.

ALGORITHM NWT (m, z, ϵ, γ)

$v \leftarrow 0$

For $i = n, \ldots, m$ do:

$$z \leftarrow z + \epsilon; \quad v \leftarrow a_i + zv; \quad \text{if } \gamma = \text{True}, a_i \leftarrow v$$

End $\blacksquare$

Then we have:

VALUE (case (i))(c)		NWT(1, c, 0, False)
VALUE (case (iii))(c)		NWT(1, $c - n$, 1, False)
TAYLOR(c)	$m = 1, \ldots, n$:	NWT(m, c, 0, True)
STIRLING	$m = 1, \ldots, n$:	NWT(m, $m - 1$, 0, True)
REV STIRLING	$m = 1, \ldots, n$:	NWT(m, $m - n - 1$, 1, True)

The FORTRAN subprogram POLY contains a parameter OPTION which determines which of the algorithms is chosen, see Table 19.1.

SUBROUTINE SPECIFICATIONS

(1) *Name of subroutine:* POLY.

(2) *Calling statement:* CALL POLY(N,A,X0,OPTION,VAL).

(3) *Purpose of subroutine:* Operations on polynomials in power and factorial form.

(4) *Descriptions of variables in calling statement:*

Name	Type	I/O/W/B	Description
N	INTEGER	I	N−1 is the degree of the input polynomial.
A	INTEGER(N)	I/O	
XO	INTEGER	I	See Table 19.1.
OPTION	INTEGER	I	
VAL	INTEGER	O	

Table 19.1

Algorithm	OPTION	XO	VAL	Input $A(1)$, ..., $A(N)$	Output $A(1)$, ..., $A(N)$
REV STIRLING	−3			Coeff. as in (3)	Coeff. as in (5)
STIRLING	−2			Coeff. as in (5)	Coeff. as in (3)
VALUE(iii)	−1	c	$f(c)$	Coeff. as in (3)	Unchanged
VALUE(i)	0	c	$f(c)$	Coeff. as in (5)	Unchanged
TAYLOR	>0	c		Coeff. as in (5)	$A(1)$, ..., A(OPTION) : Taylor coeff. at c

(5) *Other routines which are called by this are:* None.

(6) *Number of* FORTRAN *instructions:* 19.

```
SUBROUTINE POLY(N,A,XO, OPTION,VAL)
INTEGER A(N),XO, OPTION,VAL,Z,W,EPS
LOGICAL GAMMA
GAMMA=OPTION.NE.O.AND. OPTION.NE.-1
N1=MAXO(1,MINO(N,OPTION))
IF(OPTION.LT.-1)  N1=N
EPS=MOD(MAXO(-OPTION,O),2)
W=-N*EPS
IF(OPTION .GT.-2) W=W+XO
DO 1  M=1,N1
VAL=0
Z=W
DO 9  I=M,N
Z=Z+EPS
VAL=A(N+M-I)+Z*VAL
```

```
9   IF(GAMMA)   A(N+M-I)=VAL
1   IF(OPTION.LT.0)  W=W+1
    RETURN
    END
```

SAMPLE OUTPUT

POLY was called four times with $N=6$, $A(1)=\cdots=A(5)=0$, $A(6)=1$ (i.e., $f(x) = x^5$ or $f(x) = (x)_5$, depending on OPTION):

```
OPTION=6;X0=1;     Output:    1    5    10   10    5   1
OPTION=6;X0=-1;    Output:   -1    5   -10   10   -5   1
OPTION=-2;         Output:    0    1    15   25   10   1
OPTION=-3;         Output:    0   24   -50   35  -10   1
```

20

Chromatic Polynomial of a Graph (CHROMP)

Let G be a graph of n vertices. By a *proper coloring* of G we mean an assignment of colors to the vertices of G in such a way that the endpoints of no edge of G have the same color. In Fig. 20.1, we show a graph G of 4 vertices and a proper coloring of G in three colors R, Y, B.

For a fixed positive integer λ, the number of proper colorings of G *in λ or fewer* colors is denoted by $P(\lambda)$. In counting $P(\lambda)$, the vertices of G are regarded as labeled with the labels $1, 2, \ldots, n$, and two proper colorings of G are different if any one of the ordered pairs $(V, \text{color of } V)$ $(V = 1, 2, \ldots, n)$ are different in the two colorings. If G, for example, is a triangle, then $P(0) = 0$, $P(1) = 0$, $P(2) = 0$, $P(3) = 6$, $P(4) = 24, \ldots, P(\lambda) = \lambda(\lambda - 1)(\lambda - 2)$ $(\lambda \geqq 0)$.

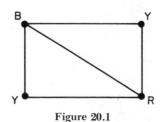

Figure 20.1

The values $P(\lambda)$ $(\lambda = 0, 1, 2, \ldots)$ turn out to be the values of a polynomial P at the nonnegative integers λ. This polynomial is called the *chromatic polynomial* of the graph G. Thus, the chromatic polynomial of the triangle is $\lambda(\lambda - 1)(\lambda - 2)$. We develop, in this chapter, a new algorithm for the computation of the coefficients of the powers of λ in the chromatic polynomial of a graph.

For a given graph G, let u and v be any two vertices which are joined by an edge, E_{uv}. Let $G - E_{uv}$ denote the new graph which is obtained from G by deleting the edge E_{uv}. Let G_{uv} denote the new graph which is obtained from G by *identifying* the two vertices u, v. This means that G_{uv} has, instead of the two vertices u and v, a single vertex uv such that uv is joined by an edge to every vertex w which was joined, in G, to u, or to v, or to both. All other vertices and edges of G_{uv} are as they were in G.

Consider the collections of all $P(\lambda; G)$ proper colorings of G, the $P(\lambda; G - E_{uv})$ proper colorings of $G - E_{uv}$, and the $P(\lambda; G_{uv})$ proper colorings of G_{uv}. Clearly, there are more colorings of $G - E_{uv}$ than of G because there is one less edge-constraint on the colorings. Indeed, the excess of $P(\lambda; G - E_{uv})$ over $P(\lambda; G)$ is exactly the number of colorings of G which are proper except that the two ends u, v of E_{uv} are given the same color. But such colorings are identical with proper colorings of G_{uv}: from any such coloring C of G_{uv}, construct a coloring C' of G in which u and v are both given the color of uv in C and all other vertices are given the colors which they had in C. It follows that

(1) $$P(\lambda; G) = P(\lambda; G - E_{uv}) - P(\lambda; G_{uv})$$

Observe that in this fundamental and well-known relation G has n vertices and E edges, $G - E_{uv}$ has n vertices and $E - 1$ edges, and G_{uv} has $n - 1$ vertices and $<E$ edges. Hence, (1) is actually a *reduction* formula. For theoretical purposes, (1) is useful because from it we can prove theorems by induction. For practical purposes, (1) can be used iteratively to compute $P(\lambda; G)$.

As an example, let us prove that $P(\lambda; G)$ is a polynomial in λ of degree n. If G has 0 edges and n vertices, then $P(\lambda; G) = \lambda^n$, so the theorem holds. If true for graphs of $<E$ edges and any number of vertices, then by (1) it remains true for graphs of E edges and any number of vertices, completing the proof. It follows that if G has n vertices, we may write $P(\lambda; G)$ in the form

(2) $$P(\lambda; G) = \lambda^n - a_{n-1}\lambda^{n-1} + a_{n-2}\lambda^{n-2} - \cdots + (-1)^{n-1}a_1\lambda$$

A similar induction will prove that $a_j \geqq 0$ $(j = 1, \ldots, n - 1)$ in (2), so that the coefficients alternate in sign.

We return now to our algorithm for computation of $a_1, \ldots, a_{n-1}$, by considering a certain *binary tree* which can be constructed from our given graph G. In case G is the graph of Fig. 20.1, the tree in question looks like Fig. 20.2.

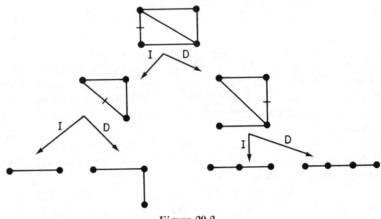

Figure 20.2

At the top of the tree is the original graph G with one of its edges marked. It has two descendants, IG and DG. IG is obtained from G by identifying the two endpoints of the marked edge. DG is obtained from G by deleting the marked edge. Next, in IG and DG, we mark an edge and repeat the process. According to Eq. (1),

$$(3) \quad P(\lambda; G) = P(\lambda; DG) - P(\lambda; IG)$$
$$= P(\lambda; DDG) - P(\lambda; IDG) - P(\lambda; DIG) + P(\lambda; IIG)$$

Why, in Fig. 20.2, did we choose to halt when we did? Because the graphs in the bottom line are all trees (connected, with no circuits). In the present context, the advantage of a tree is that *all trees on n vertices have the same chromatic polynomial*, namely

$$(4) \qquad\qquad t_n(\lambda) = \lambda(\lambda - 1)^{n-1} \quad (n \geqq 1)$$

We can prove this by induction on n: If T is a tree on n vertices, choose a terminal vertex (vertex with just one incident edge) u of T and let v be the unique other vertex of T which is connected to u. Then $T - E_{uv}$ is a tree of $n - 1$ vertices and an isolated point, which, inductively, has the chromatic polynomial

$$P(\lambda; T - E_{uv}) = \lambda \cdot (\lambda - 1)^{n-2} \cdot \lambda = \lambda^2(\lambda - 1)^{n-2}$$

On the other hand, if we identify u and v in T, we obtain a tree of $n-1$ vertices so that, inductively,

$$P(\lambda; T_{uv}) = \lambda(\lambda - 1)^{n-2}$$

Then (1) yields

$$P(\lambda; T) = \lambda^2(\lambda - 1)^{n-2} - \lambda(\lambda - 1)^{n-2} = \lambda(\lambda - 1)^{n-1}$$

as claimed.

Hence, if b_i is the number of trees on i vertices which are produced by the delete-and-identify algorithm above, then the chromatic polynomial of G is exactly

$$(5) \qquad P(\lambda; G) = \sum_{j=1}^{n} (-1)^{n-j} b_j \lambda(\lambda - 1)^{j-1}$$

As a by-product of the algorithm, notice that we have just proved the following well-known

Theorem If the chromatic polynomial of a graph is written in the form (5), then the coefficients b_j are nonnegative ($j = 1, n$).

Equation (5) is called the *Tutte polynomial* form of $P(\lambda; G)$. For example, Eq. (3) now implies that for the graph G of Fig. 20.1,

$$\begin{aligned}
P(\lambda; G) &= t_4(\lambda) - t_3(\lambda) - t_3(\lambda) + t_2(\lambda) \\
&= \lambda(\lambda - 1)^3 - 2\lambda(\lambda - 1)^2 + \lambda(\lambda - 1) \\
&= \lambda(\lambda - 1)(\lambda - 2)^2
\end{aligned}$$

Next, how can we be sure that every downward path terminates at a tree? By taking care that, at each stage, the edge which we mark has the property that its removal will not disconnect the graph. One convenient way of doing this is, after having previously found a spanning tree for the graph, to mark only edges which lie outside that spanning tree. Evidently, the removal of any such edge will not disconnect G, and, having removed all such edges, we will be left with a tree.

How can we be sure to carry out every possible combination of I's and D's? A well-known algorithm for doing this is to throw all the work to be done later onto a "stack."

(A) $G' \leftarrow G$; stack $\leftarrow$ empty stack.
(B) If G' is a tree, to (C); Stack $\leftarrow DG'$; $G' \leftarrow IG'$; To (B).
(C) Tabulate G'; If stack is empty, exit; Otherwise, $G' \leftarrow$ top graph on stack, and return to (B) ∎

The method utilizes a stack of graphs, and, as we work our way down Fig. 20.2, we travel to the left at each fork while at the same time throwing the right-hand graph DG' on top of the stack. When we reach the end of a path (G' is a tree) we take next the top graph on the stack at that time.

At each stage we call SPANFO (Chapter 18) to arrange the edges of G' in order so that the edges of a spanning tree are listed first. If G' is not a tree, then, we select for deletion the *last* edge in the list, since its removal cannot disconnect the graph. The bottom of a path, i.e., the fact that G' is a tree, is recognized simply by noting that in G' we have # edges = # vertices -1.

The whole calculation constitutes a binary tree search (e.g., Fig. 20.2) in which the tree has $\leq E$ levels, and therefore $O(2^E)$ "vertices" (i.e., graphs G'). For each G' we do $O(E)$ calculation, and so the whole algorithm requires $O(E \cdot 2^E)$ labor.

The stack of graphs is an $N \times (2 \cdot E)$ array consisting of at most N graphs whose edges are listed in $2 \times (E-1)$ or fewer locations. The last location for each graph G' on the stack is used to store $N1$, $E1$, the vertex and edge counts of G'.

In the identification process the list of endpoints of the edges is searched for the occurrence of the two vertices, u and v, to be identified, and v is replaced by u, whenever it occurs as a vertex. At the same time a notation is made for each vertex to which u or v are joined, to prevent duplicate listings. The result is the identified graph IG.

ALGORITHM CHROMP

[Input is a connected graph via its edge list; output are the numbers b_i ($i = 1, n$) of equation (5).] Note: the term "score one tree" means: set $b_m \leftarrow b_m + 1$, where m is the number of vertices of the tree at hand.

(A) [*Initialize*] $b_i \leftarrow 0$ ($i = 1, n$); Stack $\leftarrow$ empty, $G' \leftarrow G$.

(B) [*Spanfo*] Arrange edges of G' so that spanning tree comes first; if G' is a tree, score one tree and to (D); delete the last edge of G'; if this DG' is a tree, score one tree; else, write DG' on the stack.

(C) [*Identify*] Set $G' \leftarrow IG'$; if a tree, score one tree and to (D); to (B).

(D) [*Get next graph from stack*] If the stack is empty, Exit; read

$G' \leftarrow$ Stack; remove last edge from graph just read from stack; if it is a tree, score one tree and remove graph from stack; to (C) ∎

In the FORTRAN program, step **A** is performed by the instructions up to number 20, step **B** by those up to 70, step **C** by those up to 40, step **D** by those up to 100, while the remaining instructions perform additional operations on the coefficients. In step **C**, the array $A(I)$ keeps a record of vertices I that are joined to u or $v((A(I) = 1)$.

For many purposes it is desirable also to have the coefficients a_j of the form (2). These are computed by the program through a call to subroutine POLY of Chapter 19. Indeed, let $\lambda' = \lambda - 1$; then

$$P(\lambda)/\lambda = \sum_{j=1}^{n} (\lambda')^{j-1} (-1)^{n-j} b_j$$

is a polynomial which, if arranged in powers of $\lambda' + 1(=\lambda)$ gives the form (2) (after multiplying both sides again by λ). This is just what POLY does, if OPT=N.

Similarly, a call to POLY with OPT=-2 causes $P(\lambda)/\lambda$ to be written as a sum of factorial polynomials $(\lambda')_{j-1}$; this expression has the same coefficients as the expression of $P(\lambda)$ in terms of

$$\lambda, \lambda(\lambda - 1), \lambda(\lambda - 1)(\lambda - 2), \ldots, \lambda(\lambda - 1)(\lambda - 2) \ldots (\lambda - n + 1).$$

The coefficient of $\lambda(\lambda - 1) \cdots (\lambda - i + 1)$ is $1/i!$ times the number of colorings of G in *exactly* i colors.

Remark The graphs stored in STACK, at any instant, have the property that both their vertex and their edge counts are strictly decreasing. An upper estimate for the number of pairs in STACK (including the $N1$ and $E1$) is therefore

(6) $\qquad E + (E - 1) + \cdots + (E - N + 1) = NE - \dfrac{(N - 1)N}{2}$

$$= N(E - \tfrac{1}{2}N + \tfrac{1}{2})$$

SUBROUTINE SPECIFICATIONS

(1) *Name of subroutine:* CHROMP.
(2) *Calling statement:* CALL CHROMP(N,E,ENDPT,A,B,C,STACK NSTK).
(3) *Purpose of subroutine:* Calculate the chromatic polynomial of a connected graph.

(4) *Descriptions of variables in calling statement:*

Name	Type	I/O/W/B	Description
N	INTEGER	*I*	Number of vertices of G.
E	INTEGER	*I*	Number of edges of G.
ENDPT	INTEGER(2,E)	*I*	ENDPT(1,I),ENDPT(2,I) are the two ends of edge I of G(I=1,E).
A	INTEGER(N)	*O*	Coefficients in Eq. (2).
B	INTEGER(N)	*O*	Coefficients in Eq. (5).
C	INTEGER(N)	*O*	$P(\lambda) = \sum_{I=1}^{N} C(I)(\lambda)_I.$
STACK	INTEGER(2,NSTK)	W	Working storage.
NSTK	INTEGER	*I*	Maximum size of stack (see Eq. (6)).

(5) *Other routines which are called by this one:* SPANFO,POLY.
(6) *Number of Fortran instructions:* 59.
(7) *Remarks:* Input list ENDPT is destroyed.

```
      SUBROUTINE CHROMP(N,E,ENDPT,A,B,C,STACK,NSTK)
      IMPLICIT INTEGER(A-Z)
      DIMENSION ENDPT(2,E),STACK(2,NSTK),A(N),B(N),
     *  C(N),EN(2)
      IS=0
      DO 5  I=1,N
5     B(I)=0
      E1=E
      N1=N
20    CALL SPANFO(N1,E1,ENDPT,K,C)
      IF(K .NE. 1) GO TO 100
      IF(E1-N1) 40,29,30
29    B(N1)=B(N1)+1
      GO TO 70
30    DO 31  I=1,E1
      IS=IS+1
      DO 31  L=1,2
31    STACK(L,IS)=ENDPT(L,I)
32    STACK(1,IS)=N1
      STACK(2,IS)=E1-1
70    DO 71  I=1,N
71    A(I)=0
      U=MINO(ENDPT(1,E1),ENDPT(2,E1))
```

```
        V=ENDPT(1,El)+ENDPT(2,El)-U
        Ml=El-1
        El=0
        DO 75   J=1,Ml
        DO 80   L=1,2
        EN(L)=ENDPT(L,J)
        IF(EN(L) .EQ. V) EN(L)=U
80      IF(EN(L) .EQ. Nl) EN(L)=V
        DO 81  L=1,2
        IF(EN(L) .NE. U) GO TO 81
        IF(A(EN(3-L)) .NE. 0) GO TO 75
        A(EN(3-L))=1
81      CONTINUE
        El=El+1
        DO 82 L=1,2
82      ENDPT(L,El)=EN(L)
75      CONTINUE
        Nl=Nl-1
        GO TO 20
40      B(Nl)=B(Nl)+1
50      IF(IS .EQ. 0) GO TO 100
60      Nl=STACK(1,IS)
        El=STACK(2,IS)
        IS=IS-El-1
        DO 61 I=1,El
        DO 61 L=1,2
61      ENDPT(L,I)=STACK(L,IS+I)
        IF(El .EQ. Nl) GO TO 29
        IS=IS+El
        GO TO 32
100     DO 101 I=1,N
        A(I)=B(I)
101     C(I)=(1-2*MOD(N-I,2))*B(I)
        CALL POLY(N,A,1,N,V)
        CALL POLY(N,C,0,-2,V)
        RETURN
        END
```

SAMPLE OUTPUT

The chromatic polynomial of the graph shown in Fig. 20.3 was computed by CHROMP.

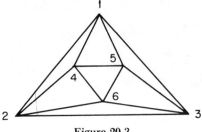

Figure 20.3

The input arrays are shown below, along with the polynomial coefficients $A(1), A(2), A(3), A(4), A(5)$, the "tree coefficients," $B(1), B(2), B(3), B(4), B(5)$, and the Stirling coefficients $C(1), C(2), C(3), C(4), C(5)$.

ENDPT(1,I):	1	1	1	1	2	2	2	3	3	4	4	5
ENDPT(2,I):	2	3	4	5	3	4	6	5	6	5	6	6

A(I):	64	154	137	58	12	1
B(I):	0	11	25	20	7	1
C(I):	0	0	1	3	3	1

The chromatic polynomial of this graph is therefore

$$
\begin{aligned}
P(\lambda) &= \lambda^6 - 12\lambda^5 + 58\lambda^4 - 137\lambda^3 + 154\lambda^2 - 64\lambda \\
&= \lambda(\lambda - 1)^5 - 7\lambda(\lambda - 1)^4 + 20\lambda(\lambda - 1)^3 \\
&\quad - 25\lambda(\lambda - 1)^2 + 11\lambda(\lambda - 1) \\
&= (\lambda)_3 + 3(\lambda)_4 + 3(\lambda)_5 + (\lambda)_6
\end{aligned}
$$

21

Composition of Power Series (POWSER)

Suppose that we are given two power series

(1) $$g(z) = b_1 z + b_2 z^2 + b_3 z^3 + \cdots$$
(2) $$h(z) = c_1 z + c_2 z^2 + c_3 z^3 + \cdots$$

and that we want to calculate the coefficients of the composite

(3) $$f(z) = g(h(z)) = a_1 z + a_2 z^2 + \cdots$$

Such problems arise frequently in combinatorics, for instance, if

(4) $$g(z) = h(z) = e^z - 1$$

the numbers $n! a_n$ $(n = 1, 2, \ldots)$ represent the number of partitions of an n-set (see Chapter 11).

As so often is the case, the most explicit method is the least desirable computationally. There is a closed formula, due to Faa di Bruno, which expresses the coefficients of $g(h(z))$ in terms of those of $g(z)$ and of $h(z)$. It states that

(5) $$a_j = \sum b_\mu \mu! \tau_\mu^{(j)} \quad (j \geq 1)$$

in which

(6) $$\tau_k^{(j)} = \sum \frac{c_1^{m_1} c_2^{m_2} \cdots}{m_1! m_2! \cdots} \quad (j \geq 1; 1 \leq k \leq j)$$

where the sum runs over the partitions Π of j into exactly k parts, and m_i is the multiplicity of i in Π. The amount of computational labor involved here is astronomical compared to that in either of the next two approaches, and so we shall not discuss (5) further.

Next, a rather neat algorithm can be constructed by the use of two linear arrays, say $\rho_1, \ldots, \rho_n; a_1, \ldots, a_n$:

(B1) $\rho_i \leftarrow c_i \ (i = 1, n); \ a_i \leftarrow 0 \ (i = 1, n); \ q \leftarrow 1$.
(B2) $a_i \leftarrow b_q \rho_i + a_i \ (i = 1, n); \ \text{If } q = n, \text{ exit}.$

(B3) $q \leftarrow q + 1; \ \rho_i \leftarrow \sum_l \rho_l c_{i-l} \ (i = 1, n); \ \text{To (B2)} \ \blacksquare$

Thus, we find the coefficients c_i of a single power of $h(z)$, the qth power, accumulate the contribution $b_q h(z)^q$ to the coefficients of $f(z)$, then go to the next power of $h(z)$, etc.

The algorithm is quite nice, but we shall not use it here. Its disadvantages relative to our final choice of a method are, first, that if we want to raise $h(z)$ to the 100th power, all lower powers would have to be calculated and, second, that the powers of h are calculated each from its predecessor so that buildup of round-off error can take place. The next method calculates each power of $h(z)$ independently of the others, and only the powers that are needed (i.e., such that $b_i \neq 0$) are computed.

First, consider the special case, where we want the coefficients $a_1, \ldots, a_n$ of $f(z)$

(7) $$f(z) = (1 + h(z))^\alpha - 1$$

if a real number α and the coefficients $c_1, \ldots, c_n$ of $h(z)$ are given. Differentiation of both sides yields

$$f'(z) = \alpha(1 + h(z))^{\alpha-1}h'(z)$$

and upon multiplying both sides by $(1 + h(z))$ we find

$$f'(z)(1 + h(z)) = \alpha(f(z) + 1)h'(z)$$

Equating the coefficients of like powers of z we obtain the recurrence

(8) $$a_1 = \alpha c_1; \quad a_j = \alpha c_j + \frac{1}{j} \sum_{i=1}^{j-1} a_i c_{j-i}(\alpha(j - i) - i) \quad (j \geq 2)$$

from which $a_1, \ldots, a_n$ can be found in $O(n^2)$ operations. This is done by POWSER if OPTION=1 (see Table 21.1).

Table 21.1

Option	1	2	3
α'	α	1	Variable
I	1	0	1
q	0	0	See Eq. (9)
s	1	1	$1/c_q$

Now, back to the first problem. Let

(9) $$h(z) = c_q z^q + c_{q+1} z^{q+1} + \cdots \quad (c_q \neq 0)$$

then the coefficients of $f(z) = g(h(z))$ can be obtained from

(10) $$f(z) = \sum_{m=1}^{\infty} b_m c_q^m z^{mq} \left(1 + \frac{c_{q+1}}{c_q} z + \frac{c_{q+2}}{c_q} z^2 + \cdots \right)^m$$

by applying POWSER with OPTION=1 to evaluate the mth power of the series in parentheses. This is OPTION=3 of POWSER (Table 21.1).
For the important special case where $g(z) = e^z - 1$:

(11) $$f(z) = \exp h(z) - 1$$

our algorithm is, of course, able to handle the problem. The calculation of n coefficients of $f(z)$ would require $O(n^3)$ operations. We can, however, calculate these n coefficients of f in $O(n^2)$ operations. Indeed, by direct logarithmic differentiation of (11) we find a recurrence formula

(12)
$$a_j = c_j + \frac{1}{j} \sum_{i=1}^{j-1} a_i c_{j-i}(j - i) \quad (j \geq 2)$$

$$a_1 = c_1$$

for the required $a_1, \ldots, a_n$. Our program POWSER provides for this case under OPTION=2.
We exploit the similarities in form between the recurrences (8) and (12) by noting that both are special cases of the recurrence

(13) $$a_j = s\left(\alpha' c_j + \frac{1}{j} \sum_{i=1}^{j-1} a_i c_{j-i+q}(\alpha'(j - i) + iI)\right) \quad (j = 1, 2, \ldots, n')$$

The parameters α', I, q, s of the general form (13) are set to the values shown in Table 21.1 in the cases of Options 1, 2, or 3.

ALGORITHM POWSER (Options 1, 2, and 3)

See Tables 21.1, 21.2.

(A) [*Entries for Options 1, 2*] $\alpha' \leftarrow (2\alpha - 1) + (1 - \alpha)*\texttt{OPTION}$;
$q \leftarrow 0$; $s \leftarrow 1$; $n' \leftarrow n$; $I \leftarrow 2 - \texttt{OPTION}$: go to (**C1**).

(B) [*Entry for Option 3*] $d_i \leftarrow b_1 c_i (i = 1, n)$; $q \leftarrow \min\{i | c_i \neq 0\}$;
$s \leftarrow 1/c_q$; $I \leftarrow 1$; Do step (**C**) for $m = 2, \ldots, \lfloor n/q \rfloor$ and then go to step (**D**).

(C) If $b_m = 0$, next m; $\alpha' \leftarrow m$; $r \leftarrow b_m c_q^{\,m}$; $n' \leftarrow n - mq$

    (**C1**) For $j = 1$, n' calculate a_j by (13); If Option = 1 or 2, then exit;

    (**C2**) $d_{mq} \leftarrow d_{mq} + r$; For $i = 1$, n' set $d_{i+mq} \leftarrow d_{i+mq} + ra_i$; next m.

(D) For $i = 1$, n: $a_i \leftarrow d_i$; Exit ■

Finally, let f, h be given power series, and consider the calculation of $g(z) = f(h^{-1}(z))$. Now, if $k(z)$ is *any* power series with $k(0) = 0$, $k'(0) \neq 0$, so k^{-1} exists, then

$$(14) \qquad f \circ k^{-1} = g \circ h \circ k^{-1}$$

In fact, we choose functions $k_1, \ldots, k_n$ such that

$$(15) \qquad h \circ k_1^{-1} \circ k_2^{-1} \circ \cdots \circ k_n^{-1}(z) = z + O(z^{n+1});$$

then the first n coefficients of $f \circ k_1^{-1} \circ \cdots \circ k_n^{-1}$ are those of g, and the problem is solved. Let

$$(16) \qquad h(z) = c_1 z + c_2 z^2 + \cdots + c_n z^n + \cdots,$$

then we choose $k_1(z) = c_1 z$, and we find

$$(17) \qquad h_1(z) = h \circ k_1^{-1}(z) = z + c_2' z^2 + c_3' z^3 + \cdots$$

where $c_i' = c_i / c_1^i$. Next, we take $k_2(z) = z + c_2' z^2$; then

$$(18) \qquad h_2(z) = h_1 \circ k_2^{-1}(z) = z + c_3'' z^3 + c_4'' z^4 + \cdots$$

and inductively, if

$$(19) \qquad h_i(z) = z + c_{i+1}^* z^{i+1} + \cdots$$

we take $k_{i+1}(z) = z + c_{i+1}^* z^{i+1}$, and obtain $h_{i+1} = h_i \circ k_{i+1}^{-1}$, of the form (19), with $i \leftarrow i + 1$, and new numerical values for the coefficients.

Table 21.2 Description of Options in POWSER Subroutine

Purpose[a]	OPTION=	Input	Output	Operation time
$f = (1+h)^\alpha - 1$	1	$n; \alpha; c_1, \ldots, c_n;$	$a_1, \ldots, a_n$	$O(n^2)$
$f = e^h - 1$	2	$n; c_1, \ldots, c_n;$	$a_1, \ldots, a_n$	$O(n^2)$
$f = g(h)$	3	$n; b_1, \ldots, b_n;$	$a_1, \ldots, a_n$	$O(n^3)$
		$c_1, \ldots, c_n;$		
$g = f(h^{-1}(z))$	4	$n; c_1, \ldots, c_n; c_1 \neq 0;$	$b_1, \ldots, b_n$	$O(n^3)$
		$a_1, \ldots, a_n;$		

[a] $f(z) = a_1 z + a_2 z^2 + \cdots$, $g(z) = b_1 z + b_2 z^2 + \cdots$, $h(z) = c_1 z + c_2 z^2 + \cdots$.

To calculate the coefficients of h_{i+1}, we merely have to expand $h_i(z)$ in powers of $(z + c_{i+1}^* z^{i+1})$; the coefficients in this expansion are then the coefficients of $h_{i+1}(z)$, because $h_i(z) = h_{i+1}(z + c_{i+1}^* z^{i+1})$. This expansion is done by a loop very similar to TAYLOR in POLY (Chapter 19). To calculate g, we merely do unto the coefficients of f what we do unto the coefficients of h.

ALGORITHM POWSER (Option 4)

See Table 21.2.

(A) For $i = 1, n$: $\{b_i \leftarrow a_i/c_1{}^i, d_i \leftarrow c_i/c_1{}^i\}$; if $n = 1$, Exit.
(B) For $m = 1, n$: $s \leftarrow -d_m$; $m_0 \leftarrow m - 1$;
   for $i = m, n$:
     for $l = i, n$: $b_l \leftarrow b_l + s b_{l-m_0}$
$$d_l \leftarrow d_l + s d_{l-m_0}$$
Exit ■

SUBROUTINE SPECIFICATIONS

(1) *Name of subroutine:* POWSER.
(2) *Calling statement:* CALL POWSER(A,B,C,N,ALPHA, OPTION,D).
(3) *Purpose of subroutine:* Compose power series.
(4) *Descriptions of variables in calling statement:*

Name	Type	I/O/W/B	Description
A	DOUBLE PRECISION(N)	I/O	$f(z) = A(1)z + A(2)z^2 + \cdots + A(N)z^N$.
B	DOUBLE PRECISION(N)	I/O	$g(z) = B(1)z + \cdots + B(N)z^N$.
C	DOUBLE PRECISION(N)	I	$h(z) = C(1)z + \cdots + C(N)z^N$.
N	INTEGER	I	Number of coefficients to be calculated.
ALPHA	DOUBLE PRECISION	I	Exponent of $1 + h(z)$ if OPTION=1.
OPTION	INTEGER	I	See Table 21.2.
D	DOUBLE PRECISION(N)	W	Working storage.

(5) *Other routines which are called by this one:*   None.

(6) *Number of* FORTRAN *instructions:*   59.

(7) *Remarks:*   All input arrays and variables (see Table 21.2) are returned unchanged. If OPTION=4, then C(1) $\neq$ 0.

```
      SUBROUTINE POWSER(A,B,C,N,ALPHA,OPTION,D)
      DOUBLE PRECISION A(N),B(N),C(N).
     * D(N),ALPHA.ALP,R,S,T,V,DFLOAT
      INTEGER Q, OPTION
      IND=1
      Q=0
      IF(OPTION-3)10,30,40
10    M1=0
      S=1.
      IF(OPTION .EQ. 2) IND=0
      ALP=1.
      IF(OPTION .EQ. 1) ALP=ALPHA
      N1=N
15    DO 11 J=1,N1
      V=0
      IF(J .EQ. 1) GO TO 11
      J1=J-1
      DO 12 I=1,J1
12    V=V+A(I)*C(J-I+Q)*(ALP*(J-I)-IND*I)
11    A(J)=(ALP*C(J)+V/DFLOAT(J))*S
      IF(OPTION-2) 43,43,36
30    DO 31 I=1,N
31    D(I)=B(1)*C(I)
      DO 33 Q=1,N
```

```
      IF(C(Q) .NE. 0) GO TO 34
33    CONTINUE
      GO TO 38
34    S=1./C(Q)
      M=1
35    M=M+1
      M1=M*Q
      IF(M1 .GT. N) GO TO 38
      IF(B(M) .EQ. 0.) GO TO 35
      ALP=M
      R=B(M)*C(Q)**M
      D(M1)=D(M1)+R
      N1=N-M1
      IF(N1)38,38,15
36    M1=M*Q
      DO 37 I=1,N1
37    D(I+M1)=D(I+M1)+A(I)*R
      GO TO 35
38    DO 39 I=1,N
39    A(I)=D(I)
      RETURN
40    T=1.
      DO 41 I=1,N
      T=T/C(1)
      B(I)=A(I)*T
41    D(I)=C(I)*T
      IF(N .EQ. 1) RETURN
      DO 42 M=2,N
      S=-D(M)
      MO=M-1
      DO 42 I=M,N
      DO 42 L=I,N
      B(L)=B(L)+S*B(L-MO)
42    D(L)=D(L)+S*D(L-MO)
43    RETURN
      END
```

FIRST SAMPLE OUTPUT, OPTION 1

With OPTION=1, $h(z) = z$, the program computes binomial coefficients. Output with $\alpha = 7$, $n = 10$ is on the next page.

```
7.0000000   21.0000000   35.0000000   35.0000000   21.0000000
7.0000000    1.0000000    0.0000000    0.0000000    0.0000000
```

SECOND SAMPLE OUTPUT, OPTION 1

If b_j is the number of binary trees on j vertices, then the generating function

$$\sum_{j=0}^{\infty} b_j z^j = \frac{1}{2z}(1 - \sqrt{1 - 4z})$$

is well known. We took OPTION=1, $h(z) = -4z$, $\alpha = 0.5$, $N = 11$, and thereby obtained $a_n = -2b_{n-1}$ $(n = 1, \ldots, 11)$ which are shown below.

```
   -2.0000000      -2.0000000      -4.0000000
  -10.0000000     -28.0000000     -84.0000000
 -264.0000000    -857.9999999   -2859.9999999
-9723.9999999  -33591.9999994
```

SAMPLE OUTPUT, OPTION 3

Let $\Phi(n)$ denote the number of ways of writing

$$n = 5x + 9y + 17z \quad (x, y, z \geq 0)$$

Then evidently

$$\frac{1}{(1 - t^5)(1 - t^9)(1 - t^{17})} = \sum_{n=0}^{\infty} \Phi(n) t^n$$

$$= \exp\left\{ \log \frac{1}{1 - t^5} + \log \frac{1}{1 - t^9} + \log \frac{1}{1 - t^{17}} \right\}$$

$$= \exp\left\{ \sum_{r=1}^{\infty} \frac{\bar{c}_r t^r}{r} \right\}$$

where $\bar{c}_r$ is the sum of those elements of the subset of $\{5, 9, 17\}$ which divide r. We can calculate $\Phi(n)$ from POWSER with input OPTION=3 and

$$b_i = \frac{1}{i!} \quad (i = 1, \ldots, n)$$

$$c_i = \frac{1}{i} \sum_{d \mid i} \{\delta_{d,5} + \delta_{d,9} + \delta_{d,17}\} d \quad (i = 1, n)$$

The output of such a calculation, with N=50, follows. The numbers printed are $\Phi(n)$ ($n = 1, 50$).

```
0.00  0.00  0.00  0.00  1.00  0.00  0.00  0.00
1.00  1.00  0.00  0.00  0.00  1.00  1.00  0.00
1.00  1.00  1.00  1.00  0.00  1.00  1.00  1.00
1.00  1.00  2.00  1.00  1.00  1.00  1.00  2.00
1.00  2.00  2.00  2.00  2.00  1.00  2.00  2.00
2.00  2.00  2.00  3.00  3.00  2.00  2.00  2.00
3.00  3.00
```

SAMPLE OUTPUT, OPTION 4

With OPTION=4, $b_i = 1/i$ ($i = 1, n$), $n = 10$, and $f(z) = z$, the program will find the coefficients $a_1, \ldots, a_{10}$ of the inverse function of $\log 1/(1 - z)$, namely, of $1 - e^{-z}$. The output follows.

```
 1.0000000  -0.5000000   0.1666667  -0.0416667   0.0083333
-0.0013889   0.0001984  -0.0000248   0.0000028  -0.0000003
```

22

Network Flows (NETFLO)

In this chapter we will consider a remarkable family of combinatorial algorithms first dealt with by Ford and Fulkerson. These are the "network flow" problems, and included as special cases are (a) finding a maximum matching of a bipartite graph, (b) discovering if a family of sets possesses a system of distinct representatives, (c) computing the Dilworth number of a partially ordered set, (d) finding the edge-connectivity or vertex-connectivity of a graph, and (e) determining if a given pair of vectors are or are not the row and column sum vectors of a matrix of zeros and ones, and if so, finding such a matrix, etc.

All of the above problems and many more can be solved with an amount of labor which is a low power of the order of complexity of the problem, i.e., the algorithm is in each case a very efficient method for handling the problem.

The general framework in which we deal with all these problems simultaneously is that of network flows. To state the problem, we need a graph, with the following additional structure:

(i) One vertex is designated *source*, another *sink*; denote them x, z, respectively. It is essential that $x \neq z$.

(ii) Associated with each edge pq are nonnegative numbers c_{pq} and c_{qp}, called the *capacities* of the edge in the directions

$p \to q$ and $q \to p$, respectively. Either c_{pq} or c_{qp} may be zero. If a pair pq is not an edge in the graph, we may assign it capacity zero in both directions. (Alternatively, we may delete all edges with only zero capacities.)

A *cut* in a network is a subset S of the vertices, such that $x \in S$ and $z \in S$. The *capacity* of the cut is then defined by

$$(1) \qquad \mathrm{cap}(S) = \sum_{\substack{p \in S \\ q \notin S}} c_{pq}$$

A *flow* in a network is a function φ defined on the edges, which satisfies Kirchoff's law. For convenience we define both φ_{pq} and φ_{qp}, and require

$$(2) \qquad \varphi_{qp} = -\varphi_{pq}$$

If $\varphi_{pq} > 0$, we consider the flow to "go" from p to q; it is *outgoing* flow for p; if $\varphi_{pq} < 0$, we have *incoming* flow to p. Kirchoff's law states the equality of these two kinds of flow at each $p \neq x, z$; that is,

$$(3) \qquad \sum_q \varphi_{pq} = 0 \qquad (p \neq x,z)$$

A flow is *permissible* if for all p, q in the network

$$(4) \qquad \varphi_{pq} \leq c_{pq}$$

The *value* of a flow is the quantity

$$(5) \qquad f(\varphi) = \sum_q \varphi_{xq}$$

If S is a cut, and φ a flow, then it is easy to see that

$$(6) \qquad f(\varphi) = \sum_{\substack{p \in S \\ q \notin S}} \varphi_{pq}$$

because all sums (3) for $p \in S$ are zero, except for $p = x$:

$$\sum_{\substack{p \in S \\ q \notin S}} \varphi_{pq} = \sum_{\substack{p \in S \\ q}} \varphi_{pq} - \sum_{\substack{p \in S \\ q \in S}} \varphi_{pq} = f(\varphi) - 0$$

the last sum being zero because $[\varphi_{pq}]_{p,q \in S}$ is a skew-symmetric matrix.

From (1), (4), and (6) we see that for any permissible flow φ and any cut S we have

$$(7) \qquad f(\varphi) \leq \mathrm{cap}(S)$$

The max-flow-min-cut theorem of Ford and Fulkerson asserts that there exist a permissible flow φ and a cut S which give equality in (7). Thus, by finding a flow φ and a cut S for which we have equality in (7), we prove the max-flow-min-cut theorem, and have in fact found a maximal flow and a minimal cut.

Ford and Fulkerson's original idea was to find a path from source to sink, every edge of which permits a positive flow. A maximum amount of flow is then pushed through this path, and the residual capacities in the edges of this path are reduced by the amount of the flow in this path. The process is repeated until termination. Let S be the set of vertices that can then be reached from the source along a flow-admitting path. Clearly, $z \notin S$, and S is a cut. It can be shown that the capacity of S equals the total flow.

Ford and Fulkerson's method was quite inefficient; in fact, if the capacities were real numbers, a maximal flow might not be obtained in a finite number of steps. Since then several drastic improvements have been made, which also use flow-pushing along paths, and have satisfactory time bounds. Recently, however, Karzanov developed a method which has the even better time bound $O(n^3)$ for a graph on n vertices, and which uses preflows.

First, we describe briefly the use of the network flow algorithm in the solution of the five problems mentioned above.

(a) Maximum Matching

Given a bipartite graph $G(S, T)$. Adjoin a source and a sink. Connect the source to each $s \in S$, and connect each $t \in T$ to the sink, using edges of capacity 1. Assign to each edge (s, t) capacity 1 (all other edges have capacity 0). All edges are directed as source $\rightarrow S \rightarrow T \rightarrow$ sink. The maximum value of a flow is then equal to the maximum number of edges in a matching. Any minimal cut defines a minimum edge-covering set of vertices.

(b) Systems of Distinct Representatives (SDR)

Given sets $S_1, \ldots, S_n$ composed of elements $x_1, \ldots, x_m$. Construct a bipartite graph by joining (S_i, x_j) if the element belongs to the set. A maximum matching of n edges is an SDR; it assigns a distinct x_i to each $S_j; x_i \in S_j$. It exists iff the capacity of a minimal cut is n. This is equivalent to $|S_{j_1}, \ldots, S_{j_k}| \geq k$ for all distinct $j_1, \ldots, j_k$.

(c) Dilworth Number

Given a partially ordered set P: $\{1, 2, \ldots, n\}$. Let $S = \{x_1, \ldots, x_n\}$, $T = \{y_1, \ldots, y_n\}$ and draw edge (x_i, y_j) if $i \alpha j$ in

P. If there are d edges in a maximum matching of this bipartite graph, then the Dilworth number of P is $n - d$, i.e., P can be covered by $n - d$ linearly ordered sets but not by fewer. A minimum cut defines an independent set in P of $n - d$ elements.

(d) *Edge-Connectivity of a Graph* (after Even and Tarjan)

To find the edge-connectivity of an undirected graph G, fix $j, 2 \leq j \leq n$. Take vertex 1 as source, j as sink, in the graph G. Give all edges unit capacity in both directions. If $\Phi(j)$ is the value of a maximum flow in this network, then the edge-connectivity is

$$k(G) = \min_{2 \leq j \leq n} \Phi(j)$$

(e) *0–1 Matrices*

Given vectors $(r_1, \ldots, r_m)$ and $(s_1, \ldots, s_n)$, a network is constructed from a source, vertices $x_1, \ldots, x_m, y_1, \ldots, y_n$, and a sink. There are edges (source, x_i) of capacity $r_i (i = 1, m)$; edges (x_i, y_j) of capacity 1 $(i = 1, m; j = 1, n)$; and edges (y_i, sink) of capacities $s_i(i = 1, n)$. A 0–1 matrix A having the given row and column sums exists if and only if the maximum flow in this network saturates all edges adjacent to the source and sink. The saturation condition holds iff $0 \leq r_i \leq n$, $0 \leq s_i \leq n$ and

$$\sum_{i=1}^{m} \min(r_i, k) \geq \sum_{i=1}^{k} s_i \qquad (k = i, \ldots, n)$$

with equality for $k = n$. If so, the matrix elements are $a_{i,j} = $ flow in edge (x_i, y_j) $(i = 1, m; j = 1, n)$.

The list of examples could go on, but in Ford and Fulkerson the interested reader will find many more. In the cases above, the fact that the algorithm produces the effects claimed is in each case a by-product of a network flow proof of, respectively, the marriage theorem, P. Hall's theorem on SDR, Dilworth's theorem, Menger's theorem, and the Gale–Ryser theorem.

The method we describe now is Karzanov's, except for some drastic refinements in the bookkeeping. The construction of a max-imal flow is performed in a number of *phases*. Each phase deals with the residual capacities left after a previous phase. Let c_{pq} be a capac-ity and φ_{pq} a flow so far constructed, then $\tilde{c}_{pq} = c_{pq} - \varphi_{pq}$ is the resid-ual capacity. If the flow is from p to q, then $\tilde{c}_{pq} < c_{pq}$, but note that $\tilde{c}_{qp} > c_{qp}$.

Each phase begins with the construction of a *KZ-net*. This is a directed subgraph, with no circuits, all whose edges have positive residual capacities. In terms of the partial order which it defines on its vertices, the source x must be its unique minimal element, and the sink z its unique maximal element. Furthermore, all directed paths from source to sink, with a positive residual capacity on each edge, of minimal length, are to lie in the KZ-net.

Once a KZ-net is constructed, a maximal flow in it is found; that is, flow is pushed through until every directed path from source to sink in the net has at least one edge on which the (residual) capacity equals the (new) flow. When this stage is reached, new residual capacities are set up in preparation for a new phase. But then every directed path from source to sink in the network, all of whose edges have positive residual capacities, will be longer than the shortest ones in the previous phase. Consequently, no more than n phases will be required.

The construction of a flow in a KZ-net uses preflows: a *preflow* φ is a function which satisfies (2) and (4), but the left side of (3) need only be nonpositive. Proceeding from source to sink along all edges, the maximum amount of preflow is pushed through. If the incoming preflow at vertex v exceeds the sum of outgoing capacities, then all outgoing edges are used to capacity, and a positive *excess* $x(v)$ remains. Otherwise, only part of the outgoing capacity is used, and the excess $x(v)$ is zero. For reasons of efficiency and simplicity an outgoing edge from a vertex is given the maximum preflow before the next edge is started. — This step is a *pushout*. To remove the excesses at the various vertices, a *pushback* is performed; that is, incoming preflow is cut back to produce a zero excess. Of course, the preflow pushed back causes excesses at the receiving vertices. If not all outgoing capacity was used there, a new pushout is performed; else, further pushback is required. When all vertices other than source and sink have zero excess, a maximal flow in the KZ-net has been found. — To assure convergence and to obtain a favorable time bound, a subtle strategy in applying pushouts and pushbacks is required. For example, after a pushback has been performed on a vertex, no further changes are to be made to any incoming or outgoing preflow.

The construction of a maximal flow in a KZ-net starts with a sequence of pushouts of preflow, first from the source (it is assumed to have infinite excess, while all other vertices have initial excess zero), then from the receiving vertices onward, in an order compatible with the partial order of the KZ-net. In later stages, similar sequences of pushouts are performed, though not starting from the source. Such a sequence is called an *advance*. An advance, once started, continues until it runs out, by reaching the sink, or by getting stuck in vertices which are unable to pass on any more preflow, or both.

After a pushout, some vertices will be balanced, i.e., have zero excess, while others have positive excesses and will, in due time, be subjected to a pushback.

An *order ideal* in a KZ-net is a set of vertices which contains with each vertex all vertices closer to the sink, with respect to the partial order.

Lemma Let φ be a preflow in a KZ-net, and I an order ideal in which every vertex, except the sink, is balanced. Then there exists a permissible flow $\bar{\varphi}$ which is equal to φ on every incoming and outgoing edge of every vertex of I.

For the proof, we define $\bar{\varphi}$ equal to φ on all edges to all vertices in I. Let v be a maximal element in the complement of I. If v is balanced, define $\bar{\varphi}$ equal φ on the edges into v; otherwise first modify φ by a pushback at v. As $I \cup \{v\}$ is an order ideal, the proof is complete, by induction on the cardinality of I.

Flow into and from a vertex of an order ideal whose vertices, except the sink, are balanced, is called *old flow*, and the algorithm makes sure it becomes part of the flow to be constructed in the KZ-net. This is achieved, after each advance, by accumulating a growing order ideal, from the sink up, and, if a new vertex is balanced, putting the incoming flow in a "safe" place, inaccessible to future pushbacks. As a result, all pushbacks at these vertices will be performed only on later flow. When a vertex, which is next taken into the order ideal, is not balanced, a pushback is performed. This will be the last action (for the present phase) on the vertex, because additional incoming flow will be stopped forcibly (the incoming edges are declared *closed*), while all outgoing flow is "old." The recipient vertices for the pushed-back preflow will, in due time, be subjected to an advance. Adding new vertices to the order ideal continues until a vertex is encountered which is a candidate for a pushout. Then the advance takes precedence again.

The strategy thus outlined gives rise to what seems an endless seesaw battle between advances toward the sink on one hand, and pushbacks and safeguarding of flow towards the source on the other. However, each resumption of advancing was preceded by at least one pushback, thereby eliminating a vertex from further action. Hence there are at most n advances before all vertices except source and sink are balanced. Then the preflow is a flow.

To show this flow is maximal, consider a path in the KZ-net from source to sink. After the first advance from the source, its very first edge is filled to capacity. We proceed to show that, at all times, after a pushout or a pushback, there is always one such edge in the path, though not necessarily the first edge. Clearly, a pushout cannot destroy equality of preflow and capacity in an edge as it causes only increases in the preflow. If no pushbacks were performed on vertices along this path, we are finished. Otherwise, let e be the edge in the path from the last vertex along it that had a pushback. Such an edge exists, since the sink certainly did not have a pushback. On this edge, then, preflow and capacity are equal, because a pushback is performed only when all outgoing capacity is used up, or outgoing edges are closed by a pushout at their endpoint—but we just eliminated this latter possibility.

To count the labor to perform one phase, it is easy to see that the construction of the KZ-net, and all the pushbacks together, require $O(E)$. Similarly, the labor involved in all advances, in filling an edge "to the last drop" and passing to the next edge, requires $O(E)$. More difficult to estimate are little drips coming in, that do not fill edges to capacity. However, during each advance every vertex has at most one such edge, so here we have an estimate of $O(n^2)$. This brings the total labor for a complete network flow construction to $n \cdot O(E + n^2) = O(n^3)$.

Now comes the detailed description of the network flow algorithm. First we give the data structures which we employ; then we discuss each step, and follow it immediately by a formal algorithm, which should remove any ambiguities. To obtain an overview of the method, it should not be necessary for the reader to spell out each algorithm.

We use the following arrays. A $2 \times E$ array ϵ will hold in $\epsilon(1, i)$ and $\epsilon(2, i)$ the endpoints of edge i. If a vertex pair p, q is listed, then also q, p must be listed. On input, the edges may be in any order. On output they will be sorted in lexicographical order. A $2 \times E$ array γ will hold in $\gamma(1, i)$ the capacity of edge i. On output, $\gamma(2, i)$ will hold the flow. Part of the intermediate calculations are performed in ϵ and γ. For example, the outgoing edges from a vertex, in a phase, will be listed before the others, while the incoming edges which hold "new" preflow, will be at the end of the edge list of the vertex. Then there are several arrays of length n, denoted a, b, c, d, and x. The last one will be used to hold the excesses, and on output will hold the flow through each vertex. All arrays will hold only integers, except x and γ, which may hold real numbers.

During most of the calculations, the entries $\epsilon(1, i)$ will be replaced by pointers, so the two opposite orientations of an edge can find each other immediately. The presence of a pointer will affect the performance of an interchange of edges, a frequently needed operation.

ALGORITHM SWAP(i, j, Option)

[Interchanges edges i and j]

(A) [*Entry if Option* = 1: *edges are cross referenced*]
$\epsilon(1, \epsilon(1, j)) \leftarrow i$; $\epsilon(1, \epsilon(1, i)) \leftarrow j$; to (**B**).
(B) [*Entry if Option* = 2: *no cross references*]
For $r = 1$ to 2 do: Interchange $\epsilon(r, i) \leftrightarrow \epsilon(r, j)$; $\gamma(r, i) \leftrightarrow \gamma(r, j)$; next r
Exit ■

Prior to the beginning of any phase we initialize $\gamma(2, i)$, sort the edges in lexicographical order (we use EXHEAP, see Chapter 15) and set pointers $a(p)$ to the beginning of the list of edges from p. Edges will be permuted again later, but only within these sublists, and the values in a will remain unchanged after this point.

ALGORITHM INIT

For $i = 1$ to E do: $\gamma(2, i) \leftarrow 0$; next i.
For $i = 1$ to n do: $a(i) \leftarrow 0$; next i.
For $i = 1$ to E do: $p \leftarrow \epsilon(1, i)$; $a(p) \leftarrow a(p) + 1$; next i; $s \leftarrow 1$;
For $i = 1$ to n do: $t \leftarrow a(i)$; $a(i) \leftarrow s$; $s \leftarrow s + t$; next i Exit ■

ALGORITHM SORT

[Sorts the edges in lexicographical order]

(A) $I \leftarrow 0$
(B) Perform EXHEAP(E, I, i, j, s).
If $I > 0$, Perform SWAP(i, j, 2); to (**B**).
If $I < 0$, $s \leftarrow \epsilon(1, i) - \epsilon(1, j)$; if $s \neq 0$, to (**B**).
else $s \leftarrow \epsilon(2, i) - \epsilon(2, j)$; to (**B**).
If $I = 0$, Exit ■

At this point the task of $\epsilon(1, i)$ has been taken over by the array a. The cross references are now entered into $\epsilon(1, i)$. The lexicographical order greatly simplifies this.

ALGORITHM XREF

[Sets cross references between opposite edges; pointers placed in $\epsilon(1, *)$]

For $i = 1$ to n do: $b(i) \leftarrow a(i)$; next i.
For $i = 1$ to E do: $p \leftarrow \epsilon(2, i)$; $\epsilon(1, i) \leftarrow b(p)$; $b(p) \leftarrow b(p) + 1$; next i; Exit ∎

The KZ-net to be constructed at the beginning of a phase will appear in storage as follows. As before, $a(p)$ points to the beginning of the list of edges from p. The array entry $b(i)$ will denote the ith vertex in an enumeration which is compatible with the partial order. The source will be $b(1)$, but the sink will be deleted, as it plays only a passive part. A pointer along b will control the computations. The vertices listed to the right of it, plus the sink, form an order ideal at all times. When the pointer is moving to the left, all vertices listed to the right of it will be balanced, and the preflow safeguarded.

The outgoing edges from vertex p, that belong to the KZ-net, will be listed in locations $a(p)$ through $c(p)$, and will carry a negative sign in $\epsilon(2, *)$. The entries in γ for these edges are modified, by reducing $\gamma(1, i)$ by $\gamma(2, i)$, and setting $\gamma(2, i)$ equal zero. Thus $\gamma(1, i)$ will be a residual capacity. Meanwhile, the reverse edge still holds the actual flow, so there is no loss of information.

The end of the list of vertices from p is reserved for those incoming edges that bring in flow that has not been safeguarded. It is empty as yet, and $d(p)$, which is to point to the beginning of it, is initialized to the location just right of the last element in the p-list.

The construction of a KZ-net is performed in two operations, first an outward drive from the source, then an inward drive from the sink. In the outward drive the vertices are scanned in a breadth-first search, starting from the source. The scanning of a vertex v means the examining of those neighbors q which can be reached by a *usable* edge i (i.e., $\gamma(1, i) > \gamma(2, i)$). If such a neighbor has not been examined before, it is placed on a sequential list in b. At this point the c-entry is given the value -1, to signify the vertex has been examined. Placement of the sink in b is avoided. When the scanning of a vertex is complete, the next vertex is taken from b. Note that the distance of a vertex from the source, measured by the shortest flow-capable path, is a monotone function on b.

If the sink was not examined in the outward drive, then no KZ-net exists, hence there is no flow-capable path from source to sink, and we go to the exit procedure, with unmodified arrays ϵ, γ, and a.

The inward drive begins by making the sink *useful* with a positive sign on its c-value. Now we follow the list in b in reverse, marking as useful only those vertices which are joined to a useful vertex by a usable edge. These usable edges are moved to the initial portion of their segment in ϵ and are marked with a negative sign in $\epsilon(2, *)$; the entry in c now points to the last one of these. Also, the entries in γ are adjusted to residual capacity and zero preflow.

ALGORITHM KZNET

[Constructs a KZ-net. If construction fails, Exit0, and ϵ, γ, and a are returned unchanged. Otherwise, Exit1 and the outgoing edges from p are listed in $\epsilon(*, i)$ with $a(p) \le i \le c(p)$; sign of $\epsilon(2, i)$ reversed. Array b holds a list of the vertices of the KZ-net in an order compatible with the partial order of the KZ-net, excluding the sink. The point just past the list of edges from p is marked in $d(p)$.]

(A) [*Initialize c, d, x; prepare for outward drive*]
For $i = 1$ to n do:
  $c(i) \leftarrow 0$; $x(i) \leftarrow 0$; if $i < n$, $d(i) \leftarrow a(i + 1)$; else, $d(i) \leftarrow E + 1$;
  next i
$x(\text{source}) \leftarrow \infty$; $r \leftarrow 0$; $w \leftarrow 1$; $b(1) \leftarrow$ source; $c(\text{source}) \leftarrow -1$.

(B) [*Read next item from b*] $r \leftarrow r + 1$; if $r > w$, to (C); $p \leftarrow b(r)$;
$f \leftarrow a(p)$; $l \leftarrow d(p) - 1$
  For $i = f$ to l do:
    $q \leftarrow \epsilon(2, i)$; $\delta \leftarrow \gamma(1, i) - \gamma(2, i)$; if $c(q) \neq 0$ or $\delta = 0$, next i
    if $q \neq$ sink, $w \leftarrow w + 1$ and $b(w) \leftarrow q$; $c(q) \leftarrow -1$; next i.

(C) [*Sink examined? Initialize backward drive*]
If $c(\text{sink}) = 0$, Exit0; $c(\text{sink}) \leftarrow 1$.

(D) [*Step backward*] $r \leftarrow r - 1$; if $r = 0$, to (F); $p \leftarrow b(r)$;
$i \leftarrow a(p) - 1$; $j \leftarrow d(p) - 1$.

(E) [*Test edges for KZ-net*] if $j = i$, $c(p) \leftarrow i \cdot \text{sgn}(i - a(p) + 1)$

and to (**D**); $q \leftarrow \epsilon(2, j)$; if $c(q) \leq 0$ or $\gamma(2, j) = \gamma(1, j)$, $j \leftarrow j - 1$
and to (**E**);
$\epsilon(2, j) \leftarrow -q$; $\gamma(1, j) \leftarrow \gamma(1, j) - \gamma(2, j)$; $\gamma(2, j) \leftarrow 0$;
$i \leftarrow i + 1$; if $i < j$, perform SWAP($i, j, 1$); to (**E**).

(**F**)   [*Compactify list on b*] $K \leftarrow 0$;
      For $r = 1$ to w do:
        if $c(b(r)) \leq 0$, next r; $K \leftarrow K + 1$; $b(K) \leftarrow b(r)$; next r; Exit1 ■

The pushout step takes the excess at a vertex, and pushes it down the outgoing edges, increasing both the preflow, and the excesses at the receiving vertices. No flow is pushed down edges whose terminal points have negative x-values indicating that the terminal point had a pushback, and that the incoming edges are closed. There is a parameter P which is set to 1 if actual flow is pushed down.

ALGORITHM PUSHOUT(p, P)

[Pushes down preflow from vertex p on outgoing edges that are not closed. P is set to 1 if actual preflow is pushed through; else, P is left unchanged.]

(**A**)   [*Initialize*] $i \leftarrow c(p) + 1$.
(**B**)   [*Push preflow down one edge*] $i \leftarrow i - 1$; if $i < a(p)$, to (**D**);
      $q \leftarrow -\epsilon(2, i)$; if $x(q) < 0$, to (**B**);
      $\delta \leftarrow \min(x(p), \gamma(1, i) - \gamma(2, i))$; $\gamma(2, i) \leftarrow \gamma(2, i) + \delta$;
      $x(p) \leftarrow x(p) - \delta$; $x(q) \leftarrow x(q) + \delta$; $P \leftarrow 1$; $j \leftarrow \epsilon(1, i)$;
      if $j \geq d(q)$, to (**C**); $d(q) \leftarrow d(q) - 1$; if $j \neq d(q)$, perform SWAP
      ($j, d(q), 1$).
(**C**)   [*Any preflow left; edge saturated?*] If $x(p) > 0$, to (**B**); if
      $\gamma(1, i) = \gamma(2, i)$, $i \leftarrow i - 1$.
(**D**)   [*Update c(p)*] $c(p) \leftarrow i$; Exit ■

The safeguarding of old preflow is effected by adding it to the flow already accumulated during previous phases. Let i be an incoming edge to p, so $i \geq d(p)$; then the preflow is located in $\gamma(2, j)$ where $j = \epsilon(1, i)$ is the opposite edge, and the flow, with reverse sign, is in $\gamma(2, i)$. While moving the preflow, also the residual capacity $\gamma(1, j)$ must be adjusted.

ALGORITHM OLDFLOW(p)

[Moves old preflow to accumulated flow from previous phases.]

(A) *[Initialize]* $l \leftarrow E + 1$; if $p < n$, $l \leftarrow a(p + 1)$; $i \leftarrow d(p)$; $d(p) \leftarrow l$.

(B) *[Move preflow]* if $i = l$, Exit;
$j \leftarrow \epsilon(1, i)$; $\delta \leftarrow \gamma(2, j)$; $\gamma(2, j) \leftarrow 0$;
$\gamma(1, j) \leftarrow \gamma(1, j) - \delta$; $\gamma(2, i) \leftarrow \gamma(2, i) - \delta$; $i \leftarrow i + 1$; to (B) ∎

A pushback at p consists in reducing incoming flow in some of the edges at the end of the list of p until the excess is zero. Then we set $x(p) \leftarrow -1$ as a warning for attempts at entering further preflow. The outgoing preflow from p is incoming flow to vertices in the order ideal of vertices listed after p in b, all of whose vertices have zero excess and whose incoming flow has been safeguarded. All activity in p is frozen from now on.

ALGORITHM PUSHBACK(p)

[Pushes back excess new flow into incoming edges; then $x(p)$ set to -1.

(A) *[Initialize]* $i \leftarrow d(p)$.

(B) *[Push back preflow along edge i]* $j \leftarrow \epsilon(1, i)$; $\delta \leftarrow \min(x(p), \gamma(2, j))$; $\gamma(2, j) \leftarrow \gamma(2, j) - \delta$; $x(p) \leftarrow x(p) - \delta$; $q \leftarrow \epsilon(2, i)$; $x(q) \leftarrow x(q) + \delta$; if $x(p) > 0$, $i \leftarrow i + 1$; and to (B); $x(p) \leftarrow -1$; Exit ∎

The algorithm which controls the action of the three preceding algorithms is now surprisingly simple. There are two basic parameters, m and P. The former is a pointer, which moves back and forth along the array b. When m moves right, pushouts are performed on the vertices that need it, until the end of the list. Then a retreat takes place, during which vertices, according to their needs, are left alone, or have old flow removed, or undergo a pushback. The reversal from retreat to advance is controlled by the parameter P, which is set to 1 when a pushout moves a positive amount of preflow, and is left unchanged if no preflow is moved. The following flow chart illustrates the basic logic. When a retreat reaches the source, the phase is finished, except for a readjustment of the γ's.

FLOW CHART PREFLOW

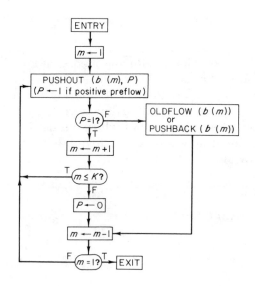

ALGORITHM PREFLOW

(**A**) [*Initialize*] $m \leftarrow 1$; $P \leftarrow 1$.

(**B**) [*Advance*] $p \leftarrow b(m)$; if $x(p) > 0$, to (**C**); else, to (**D**).

(**C**) [*Pushout*] Perform PUSHOUT(p, P); if $P = 0$, to (**G**).

(**D**) [*Next step in advance*] $m \leftarrow m + 1$; if $m \leq K$, to (**B**).

(**E**) [*Initialize for retreat*] $P \leftarrow 0$.

(**F**) [*Next step in retreat*] $m \leftarrow m - 1$; if $m = 1$, to (**H**); else, $p \leftarrow b(m)$;

      If $x(p) < 0$, to (**F**)

      If $x(p) = 0$, perform OLDFLOW(p); to (**F**)

      If $x(p) > 0$, if $a(p) \leq c(p)$, to (**C**); else, to (**G**).

(**G**) [*Pushback*] Perform PUSHBACK(p); to (**F**).

(**H**) [*Adjust ϵ and γ*] For $i = 1$ to E do:

$q \leftarrow -\epsilon(2, i)$; if $q < 0$, next i
$\epsilon(2, i) \leftarrow q$; $j \leftarrow \epsilon(1, i)$;
$\gamma(1, i) \leftarrow \gamma(1, i) - \gamma(2, j)$; $\delta \leftarrow \gamma(2, i) - \gamma(2, j)$;
$\gamma(2, i) \leftarrow \delta$; $\gamma(2, j) \leftarrow -\delta$; next i; Exit ■

The total algorithm consists of little more than calls to the various algorithms. Only minor chores are left, including the removal of the cross references.

ALGORITHM NETFLO(n, E, ϵ, γ, source, sink, a, b, c, d, x)

[Finds a maximal flow and a minimal cut in a network. Maximal flow will be in $\gamma(2, *)$; minimal cut in c. The flows through the vertices are stored in x; the flow value in x(source) and x(sink). The edge list is returned in lexicographical order, with $a(p)$ pointing to the beginning of the edge list of p.]

(A) Perform INIT, SORT, XREF.
(B) Perform KZNET; if Exit0, to (C); perform PREFLOW; to (B).
(C) For $i = 1$ to n do: $x(i) \leftarrow 0$; $c(i) \leftarrow -c(i)$; next i
     For $i = 1$ to E do: $p \leftarrow \epsilon(2, \epsilon(1, i))$; $\epsilon(1, i) \leftarrow p$;
       $x(p) \leftarrow x(p) + \max(0, \gamma(2, i))$; next i;
     x(sink) $\leftarrow x$(source); perform SORT; Exit ■

SUBROUTINE SPECIFICATIONS

(1) *Name of subroutine:* NETFLO.
(2) *Calling statement:* CALL NETFLO(N, E, ENDPT, CAPFLO,
     SOURCE, SINK, CUT, X, A, AUX1, AUX2)
(3) *Purpose of subroutine:* Find maximum flow in a network.
(4) *Description of variables in calling statement:*

Name	Type	I/O/W/B	Description
N	INTEGER	I	Number of vertices.
E	INTEGER	I	Number of edges (see (7) below).
ENDPT	INTEGER(2,E)	I	In column J are the endpoints of edge J.
CAPFLO	INTEGER(2,E)	I/O	Capacities and flows, see (7) below.
SOURCE	INTEGER	I	The designated source vertex.
SINK	INTEGER	I	The designated sink vertex.

(*continued on following page*)

CUT	INTEGER(N)	O	CUT(I)=1 if I is in a minimal cut set, otherwise O.
X	INTEGER(N)	O	X(I) is the quantity which flows through vertex I.
A	INTEGER(N)	W	A(I) points to beginning of edges from I in ENDPT on output.
AUX1	INTEGER(N)	W	Working storage.
AUX2	INTEGER(N)	W	Working storage.

(5) *Other routines which are called by this one:*   EXHEAP.

(6) *Number of* FORTRAN *instructions:*   169

(7) *Remarks:* Required input data are N,E,ENDPT,CAPFLO (1, *),SOURCE,SINK. Any edge pq is to be listed in ENDPT both as (p, q) and (q, p), with capacities in CAPFLO(1, *). On output, flows in CAPFLO(2, *), and edges sorted lexicographically.

```
      SUBROUTINE NETFLO(N,E,ENDPT,CAPFLO,SOURCE,
     *SINK,CUT,X,A,AUX1,AUX2)
      IMPLICIT INTEGER(A-Z)
      DIMENSION ENDPT(2,E),CUT(N),A(N),AUX1(N),
     *AUX2(N)
      INTEGER CAPFLO(2,E),X(N),DEL
C     CHANGE ABOVE TYPE DECLARATION FOR REAL FLOWS
C
C     START INIT
C
      DO 11 I=1,N
11    A(I)=0
      DEL=0
      DO 12 I=1,E
      CAPFLO(2,I)=0
      P=ENDPT(1,I)
      IF(P .EQ. SOURCE) DEL=DEL+CAPFLO(1,I)
12    A(P)=A(P)+1
      X(SOURCE)=DEL
      S=1
      DO 13 I=1,N
      T=A(I)
      A(I)=S
      AUX1(I)=S
```

```
13      S=S+T
        SRTRTN=0
C
C       END INIT--START SORT
C
20      INDEX=0
21      CALL EXHEAP(E,INDEX,EN1,EN2,S)
        IF(INDEX) 22,23,31
C       31 IS CALL TO SWAP-THEN TO 21
22      S=ENDPT(1,EN1)-ENDPT(1,EN2)
        IF(S .EQ. 0) S=ENDPT(2,EN1)-ENDPT(2,EN2)
        GO TO 21
23      IF(SRTRTN) 40,40,83
C
C       END SORT--START SWAP
C
30      ENDPT(1,ENDPT(1,EN1))=EN2
        ENDPT(1,ENDPT(1,EN2))=EN1
31      DO 32 R=1,2
        S=ENDPT(R,EN1)
        ENDPT(R,EN1)=ENDPT(R,EN2)
        ENDPT(R,EN2)=S
        DEL=CAPFLO(R,EN1)
        CAPFLO(R,EN1)=CAPFLO(R,EN2)
32      CAPFLO(R,EN2)=DEL
        IF(INDEX) 92,57,21
C
C       END SWAP--START XREF
C
40      DO 41 I=1,E
        Q=ENDPT(2,I)
        ENDPT(1,I)=AUX1(Q)
41      AUX1(Q)=AUX1(Q)+1
C
C       END XREF--START KZNET
C
50      INDEX=0
        DO 51 I=1,N
        IF(I .NE. SOURCE)X(I)=0
        AUX2(I)=E+1
        IF(I .LT. N) AUX2(I)=A(I+1)
```

```
51    CUT(I)=0
      READ=0
      WRITE=1
      AUX1(1)=SOURCE
      CUT(SOURCE)=-1
52    READ=READ+1
      IF(READ .GT. WRITE) GO TO 55
      P=AUX1(READ)
      LST=AUX2(P)-1
      I=A(P)-1
53    I=I+1
      IF(I .GT. LST) GO TO 52
      Q=ENDPT(2,I)
      DEL=CAPFLO(1,I)-CAPFLO(2,I)
      IF(CUT(Q) .NE. 0 .OR. DEL .EQ. 0) GO TO 53
      IF(Q .EQ. SINK) GO TO 54
      WRITE=WRITE+1
      AUX1(WRITE)=Q
54    CUT(Q)=-1
      GO TO 53
55    IF(CUT(SINK) .EQ. 0) GO TO 80
C     80 IS CALL TO EXIT PROCEDURE
      CUT(SINK)=1
56    READ=READ-1
      IF(READ .EQ. 0) GO TO 60
      P=AUX1(READ)
      EN1=A(P)-1
      EN2=AUX2(P)-1
57    IF(EN1 .EQ. EN2) GO TO 59
      Q=ENDPT(2, EN2)
      IF(CUT(Q) .GT. 0 .AND. CAPFLO(1,EN2)
     *.NE. CAPFLO(2,EN2)) GO TO 58
      EN2=EN2-1
      GO TO 57
58    ENDPT(2,EN2)=-Q
      CAPFLO(1,EN2)=CAPFLO(1,EN2)-CAPFLO(2,EN2)
      CAPFLO(2,EN2)=0
      EN1=EN1+1
      IF(EN1 .LT. EN2) GO TO 30
C     30 IS CALL TO SWAP THEN TO 57
59    IF(EN1 .GE. A(P)) CUT(P)=EN1
      GO TO 56
```

```
60    KZ=0
      DO 61 R=1,WRITE
      IF(CUT(AUX1(R)).LE. 0) GO TO 61
      KZ=KZ+1
      AUX1(KZ)=AUX1(R)
61    CONTINUE
C
C     END KZNET--START PREFLOW
C
      INDEX=-1
      M=1
71    P=AUX1(M)
      IF(X(P).GT. 0) GO TO 90
C     90 IS CALL TO PUSHOUT-THEN TO 72 OR 110
72    M=M+1
      IF(M .LE. KZ) GO TO 71
      PARM=0
73    M=M-1
      IF(M .EQ. 1) GO TO 75
      P=AUX1(M)
      IF(X(P)) 73, 100, 74
      100 IS CALL TO OLDFLOW-THEN TO 73
74    IF(A(P)-CUT(P)) 90, 90, 110
C     90 IS CALL TO PUSHOUT-THEN TO 72 OR 110
C     110 IS CALL TO PUSHBACK-THEN TO 73
75    DO 76 I=1,E
      Q=-ENDPT(2,I)
      IF(Q .LT. 0) GO TO 76
      ENDPT(2,I)=Q
      J=ENDPT(1,I)
      CAPFLO(1,I)=CAPFLO(1,I)-CAPFLO(2,J)
      DEL=CAPFLO(2,I)-CAPFLO(2,J)
      CAPFLO(2,I)=DEL
      CAPFLO(2,J)=-DEL
76    CONTINUE
      GO TO 50
C
C     END PREFLOW--START EXIT ROUTINE
C
80    DO 81 I=1,N
81    CUT(I)=-CUT(I)
      DO 82 I=1,E
```

```
        P=ENDPT(2,ENDPT(1,I))
        IF(CAPFLO(2,I) .LT. 0) X(P)=X(P)-CAPFLO(2,I)
82      ENDPT(1,I)=P
        X(SOURCE)=X(SINK)
        SRTRTN=1
        GO TO 20
C       CALL TO SORT-THEN RETURN
83      RETURN
C
C       END EXIT ROUTINE--START PUSHOUT
C
90      I=CUT(P)+1
91      I=I-1
        IF(I .LT. A(P)) GO TO 94
        Q=-ENDPT(2,I)
        IF(X(Q) .LT. 0) GO TO 91
        DEL=CAPFLO(1,I)-CAPFLO(2,I)
        IF(X(P) .LT. DEL) DEL=X(P)
        CAPFLO(2,I)=CAPFLO(2,I)+DEL
        X(P)=X(P)-DEL
        X(Q)=X(Q)+DEL
        PARM=1
        EN1=ENDPT(1,I)
        EN2=AUX2(Q)-1
        IF(EN1-EN2) 30, 92, 93
C       30 IS CALL TO SWAP-THEN TO 92
92      AUX2(Q)=EN2
93      IF(X(P) .GT. 0) GO TO 91
        IF(CAPFLO(1,I) .EQ. CAPFLO(2,I)) I=I-1
94      CUT(P)=I
        IF(PARM) 72, 110, 72
C
C       END PUSHOUT--START OLDFLOW
C
100     LST=E+1
        IF(P .LT. N) LST=A(P+1)
        I=AUX2(P)
        AUX2(P)=LST
101     IF(I .EQ. LST) GO TO 73
        J=ENDPT(1,I)
        DEL=CAPFLO(2,J)
        CAPFLO(2,J)=0
        CAPFLO(1,J)=CAPFLO(1,J)-DEL
```

```
      CAPFLO(2,I)=CAPFLO(2,I)-DEL
      I=I+1
      GO TO 101
C
C     END OLDFLOW--START PUSHBACK
C
110   I=AUX2(P)
111   J=ENDPT(1,I)
      DEL=CAPFLO(2,J)
      IF(X(P) .LT. DEL) DEL=X(P)
      CAPFLO(2,J)=CAPFLO(2,J)-DEL
      X(P)=X(P)-DEL
      Q=ENDPT(2,I)
      X(Q)=X(Q)+DEL
      I=I+1
      IF(X(P) .GT. 0) GO TO 111
      X(P)=-1
      GO TO 73
C
C     END PUSHBACK
C
      END
```

SAMPLE OUTPUT

The program NETFLO was called to solve the network shown in Fig. 22.1, in which edge capacities are shown in bold type.

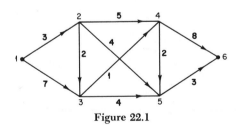

Figure 22.1

On input, the ENDPT array was

```
1 2 1 3 2 3 2 4 2 5 3 4 3 5 4 5 4 6 5 6
2 1 3 1 3 2 4 2 5 2 4 3 5 3 5 4 6 4 6 5
```

and the CAPFLO array was

3 0 7 0 2 0 5 0 4 0 1 0 4 0 2 0 8 0 3 0

- - - - - - - - - - - - - - - - - - - -

On output, the ENDPT array was

1 1 2 2 2 2 3 3 3 3 4 4 4 4 5 5 5 5 6 6
2 3 1 3 4 5 1 2 4 5 2 3 5 6 2 3 4 6 4 5

and the CAPFLO array was

3 7 0 2 5 4 0 0 1 4 0 0 2 8 0 0 0 3 0 0
3 4 −3 0 3 0 −4 0 1 3 −3 −1 0 4 0 −3 0 3 −4 −3

The CUT array was

1 0 1 0 1 0

which describes a capacity of 7, and the X array was

7 3 4 4 3 7

indicating a flow value of 7.

23

The Permanent Function (PERMAN)

Let

(1) $$A = (a_{ij})_{i,j=1}^{n}$$

be an $n \times n$ square matrix. As is well known, the *determinant* of A is the number

(2) $$\det(A) = \sum_{\sigma} \text{sgn}(\sigma) a_{1,\sigma(1)} a_{2,\sigma(2)} \cdots a_{n,\sigma(n)}$$

where the sum is over all permutations σ of $\{1, 2, \ldots, n\}$. We contrast this with the *permanent* of A, which is instead,

(3) $$\text{per}(A) = \sum_{\sigma} a_{1,\sigma(1)} a_{2,\sigma(2)} \cdots a_{n,\sigma(n)}$$

The distinction lies solely in the omission of the $\pm$ sign of the permutation σ. This omission causes the permanent not to share many of the nice properties of determinants. For instance, we have generally,

(4) $$\text{per}(AB) \neq \text{per}(A) \, \text{per}(B)$$

The permanent is of considerable combinatorial importance, however. If A, for instance, is a matrix all of whose entries are $+1$, then from (3), per $A = n!$ is the number of permutations of n letters. More generally, let A be a matrix all of whose entries are either 0 or 1. Then in (3) each term is 0 or 1, and the value of per A is just the number of permutations σ which hit only 1's in the matrix A. This fact can be exploited for various combinatorial purposes.

Let A_n be given by

$$(5) \qquad a_{ij} = \begin{cases} 0 & \text{if } i = j \\ 1 & \text{if } i \neq j \end{cases} \quad (i, j = 1, \ldots, n)$$

Then, a term in the sum (3) is 1 if and only if

$$a_{1, \sigma(1)} = 1 \quad \text{and} \quad a_{2, \sigma(2)} = 1 \quad \text{and} \quad \cdots \quad \text{and} \quad a_{n, \sigma(n)} = 1$$

which in the case (5) means that

$$\sigma(1) \neq 1 \quad \text{and} \quad \sigma(2) \neq 2 \quad \text{and} \quad \cdots \quad \text{and} \quad \sigma(n) \neq n$$

which, in turn, is so if and only if the permutation σ leaves no letter fixed. Hence, for the matrix A_n of (5),

$$(6) \qquad D_n = \text{per}(A_n)$$

is the number of fixed-point free permutations of n letters. These "rencontres" numbers are well known to be given by

$$(7) \qquad D_n = n! \left\{ 1 - 1 + \frac{1}{2!} - \frac{1}{3!} + \cdots + \frac{(-1)^n}{n!} \right\}$$

Again, let $S_1, \ldots, S_n$ be a collection of sets whose union U consists of objects $a_1, a_2, \ldots, a_n$. By a *system of distinct representatives* of the collection $S_1, \ldots, S_n$ we mean a list $y_1, y_2, \ldots, y_n$ of all of the objects in U arranged in sequence so that $y_i \in S_i$ $(i = 1, \ldots, n)$. The permanent function counts systems of distinct representatives: Let a matrix A be defined by $a_{ij} = 1$ if $a_i \in S_j$ and $a_{ij} = 0$ otherwise. This matrix A is called the *incidence matrix* of the objects in the sets. A single term in the sum (3) is $=1$ or 0 according to whether $a_{\sigma(1)}, \ldots, a_{\sigma(n)}$ is or is not, respectively, a system of distinct representatives for the sets $S_1, \ldots, S_n$. The permanent of A is therefore equal to the number of such systems.

As another application of the permanent function, by an $r \times n$ *Latin rectangle* we mean a rectangular array of r rows and n columns, whose entries are letters chosen from $\{1, 2, \ldots, n\}$ such that (a) the entries in each row constitute a permutation of $\{1, 2, \ldots, n\}$ *and* (b) the entries in each column are all different. We show a 3×5 Latin rectangle in Fig. 23.1. Our question is: In how many ways can we adjoin a new row to a given $r \times n$ Latin rectangle in such a way that the result is an $(r + 1) \times n$ Latin rec-

I	2	3	4	5
3	I	5	2	4
2	4	I	5	3

Figure 23.1

tangle? To see this question as one about permanents, define n sets $S_1, \ldots, S_n$ as follows: $i \in S_j$ if i does not appear in column j of the given Latin rectangle ($i = 1, \ldots, n; j = 1, \ldots, n$). Hence, S_j is the set of letters which might possibly appear in the jth column of any new row which is adjoined.

It is easy to check that $|S_j| = n - r$ ($j = 1, \ldots, n$), and that each letter i appears in exactly $n - r$ of the S_j's. As above, let A be the incidence matrix of the objects $1, 2, \ldots, n$ in the sets $S_1, \ldots, S_n$. Then A is an $n \times n$ matrix of 0's and 1's which has exactly $n - r$ 1's in each row and column. The permanent of A is equal to the number of systems of distinct representatives of the family of sets $S_1, \ldots, S_n$, and each system of distinct representatives is a way of adjoining a new row. Hence, per(A) is equal to the number of possible extensions. This idea can be used, for example, to prove that an extension is *always* possible. This is so because the permanent of a matrix of nonnegative entries, with constant nonzero row and column sums, is always strictly positive. Thus per $A > 0$ and an extension always exists.

Because of such applications as the above, great interest attaches to the question of estimating the size of the permanent of a square nonnegative matrix in terms of its row and column sums. First, if the entries of A are 0's and 1's, then a theorem of Bregman, conjectured by Ryser and Minc holds that

(8) $$\mathrm{per}(A) \leq \prod_{j=1}^{n} (r_j!)^{1/r_j}$$

where r_j is the number of 1's in the jth row of A.

In the other direction, if $a_{ij} \geq 0$ ($i, j = 1, \ldots, n$) and

$$\sum_{j=1}^{n} a_{ij} = 1 \quad (i = 1, \ldots, n)$$

(10)

$$\sum_{i=1}^{n} a_{ij} = 1 \quad (j = 1, \ldots, n)$$

then an unproven conjecture of van der Waerden states that

$$(11) \qquad \qquad \mathrm{per}(A) \geqq \frac{n!}{n^n}$$

which, if true, would also be the best possible.

Observe, for instance, that since (8) is known to be true, then by the construction outlined above we would have at *most*

$$(n - r)!^{n/(n-r)}$$

extensions of an r-rowed Latin rectangle to an $(r + 1)$-rowed one. On the other hand, if (11) were true, then by dividing our incidence matrix by $n - r$, we would satisfy (10) and thereby learn that a Latin rectangle has at *least*

$$(n - r)^n n!/n^n$$

extensions to one of order $r + 1$. The number L_n of $n \times n$ Latin *squares* would then satisfy

$$(12) \qquad \qquad \left\{ \prod_{\nu=1}^{n-1} \nu!^{1/\nu} \right\}^n \geqq L_n \geqq \frac{n!^{2n-1}}{n^{n^2}}$$

which would be an improvement over known bounds.

COMPUTATION OF THE PERMANENT FUNCTION

Observe that a direct application of (3) to an $n \times n$ matrix would require about $n \cdot n!$ operations for the calculation of per(A). We reduce this labor in three steps.

(1) A method due to Ryser evaluates per(A) in about $n^2 2^{n-1}$ operations. It requires an average of $n^2/2$ calculations for each of the 2^n subsets of $\{1, 2, \ldots, n\}$. Ryser's method is derived below and appears in Eqs. (17) and (18).

(2) We reduce the above by a factor of 2 by a method which makes it necessary to process only the subsets of $\{1, 2, \ldots, n - 1\}$. Thus in Eq. (24) we do about $n^2/2$ calculations for each of the 2^{n-1} subsets of $\{1, 2, \ldots, n - 1\}$, or $(n^2/4) 2^n$ operations altogether.

(3) A further reduction by a factor of $n/2$ is accomplished by arranging the sequence of subsets so as to follow a Hamilton walk on an $(n - 1)$-cube (see Chapter 1). If this is done, only a slight change in the calculation for a set S will yield the result of the calculation for the successor of S. In this way, our final algorithm PERMAN

does about n calculations for each of the 2^{n-1} subsets of $\{1, 2, \ldots, n-1\}$, for a total of $n2^{n-1}$ operations (multiplication and addition) to compute the permanent of an $n \times n$ matrix.

We first describe Ryser's formula. Suppose, for the moment, that A is an $n \times m$ matrix; let $f: \{1, \ldots, n\} \to \{1, \ldots, m\}$ denote a mapping, and let the *weight* of f be

$$(13) \qquad w(f) = \prod_{i=1}^{n} a_{i,f(i)}$$

for each of the n^m mappings f. Then we can define the permanent of such a rectangular matrix by

$$(14) \qquad \text{per}(A) = \sum_{f}{}' w(f)$$

where the prime indicates that the sum is over only whose f such that

$$\forall j \in \{1, \ldots, m\} : f^{-1}(j) \neq \varnothing$$

Note that if $m > n$, $\text{per}(A) = 0$.

Now consider the n^m objects as our mappings f, and let the jth property of one of these objects be

$$P_j : f^{-1}(j) = \phi \quad (j = 1, \ldots, m)$$

Then by the principle of inclusion–exclusion, the total weight of those objects which have none of the properties is

$$(15) \qquad \text{per } A = \sum_{T} (-1)^{|T|} N(T)$$

where T runs over all subsets of properties, and $N(T)$ is the weight of those objects which have the set T of properties. But the weight $N(T)$ of those objects is

$$(16) \qquad N(T) = \prod_{i=1}^{n} \left\{ \sum_{j \in T^c} a_{ij} \right\}$$

since the weight of every mapping with properties T occurs once in the expansion of the right-hand side. If we substitute (16) into (15), we find

$$\text{per}(A) = \sum_{T} (-1)^{|T|} \prod_{i=1}^{n} \left\{ \sum_{j \in T^c} a_{ij} \right\}$$

and, finally, if we sum over $S = T^c$, instead of T,

(17) $$\text{per } A = (-1)^n \sum_S (-1)^{|S|} \sigma_S$$

where

(18) $$\sigma_S = \prod_{i=1}^n \sum_{j \in S} a_{ij}$$

(17) is Ryser's formula. We observe that (17) requires about $2^n \cdot (n^2/2)$ operations for the computation of the permanent.

A variation which saves half of the labor will now be described. Suppose A is $n \times n$. Adjoin to A a new column $[x_1, x_2, \ldots, x_n]^T$, obtaining an $n \times (n+1)$ matrix A', and number the columns $0, 1, 2, \ldots, n$. If we apply (17) to A', we obtain

(19) $$0 = -\text{per } A' = (-1)^n \sum_{S'} (-1)^{|S'|} \sigma_{S'} + (-1)^n \sum_{S''} (-1)^{|S''|} \sigma_{S''}$$

where S' and S'' run, respectively, through all the subsets of $\{0, 1, \ldots, n\}$ which do (respectively, do not) contain 0. Hence

(20) $$0 = (-1)^n \sum_{S'} (-1)^{|S'|} \sigma_{S'} + \text{per } A$$

But

(21) $$\sigma_{S'} = \prod_{i=1}^n \left\{ x_i + \sum a_{ij} \right\}$$

Let $(S')^c$ be the complementary set of $(S' - \{0\})$. Then

(22) $$\sigma_{(S')^c} = \prod_{i=1}^n \left[x_i + \left(\sum_{j=1}^n a_{ij} \right) - \sum_{\substack{j \neq 0 \\ j \in S'}} a_{ij} \right]$$
$$= \prod_{i=1}^n \left[x_i + r_i - \sum_{\substack{j \neq 0 \\ j \in S'}} a_{ij} \right]$$

where r_i is the ith-row sum of A. Now choose $x_i = -\frac{1}{2} r_i$. Then

(23) $$\sigma_{(S')^c} = (-1)^n \sigma_{S'}$$

and the contributions of S' and $(S')^c$ to the sum in (20) are equal. We can, for example, compute only the terms in the sum (20) corresponding to S' which contain both 0 and n, and double the result.

Ryser's method, together with this modification, can be summarized as

(24a) $$x_i = a_{i,n} - \frac{1}{2} \sum_{j=1}^{n} a_{ij} \quad (i = 1, \dots, n)$$

(24b) $$\operatorname{per}(A) = (-1)^{n-1} 2 \sum_{S}{}'' (-1)^{|S|} \prod_{i=1}^{n} \left\{ x_i + \sum_{j \in S} a_{ij} \right\}$$

where S runs only over the subsets of $1, 2, \dots, n-1$.

To save our final factor of $n/2$ in the amount of computation required, observe that for each subset $S \subseteq \{1, 2, \dots, n-1\}$ we have to calculate

(25) $$f(S) = \prod_{i=1}^{n} \lambda_i(S)$$

where

(26) $$\lambda_i(S) = x_i + \sum_{j \in S} a_{ij} \quad (i = 1, \dots, n)$$

Suppose that our current subset S differs from its predecessor S' by a single element, j. Then

(27) $$\lambda_i(S) = \lambda_i(S') \pm a_{ij} \quad (i = 1, \dots, n)$$

Thus, instead of requiring $n(|S| + 1)$ operations to compute $\lambda_1, \dots, \lambda_n$ in (26), we can find them in just n operations by (27). The key to the saving is, then, generating the subsets of $1, 2, \dots, n-1$ in such a sequence that each set S differs from its predecessor only in a single element. In Chapter 1, this question was discussed in detail, and we produced Algorithm NEXSUB for doing the generation of the subsets. Hence in our present problem, Algorithm PERMAN will simply call NEXSUB to get its next subset of $1, \dots, n-1$, calculate the λ_i as in (27), $f(S)$ as in (25), and per A from (24). The program is very short, just 26 instructions.

The question of significant digits merits some attention. It is characteristic of inclusion–exclusion calculations that the terms get larger for a while (as $|S|$ increases) then smaller, that there is a good deal of cancellation between terms, and that the final answer may be much smaller than many of the individual terms in the sum. It is tempting to consider using integer arithmetic when calculating with an integer matrix. Yet, in the present situation, one may find that even though $\operatorname{per}(A)$ is small enough to fit comfortably into an integer word, intermediate quantities in the calculation may overflow. For these reasons our program is in double-precision mode.

ALGORITHM PERMAN

(A) $p \leftarrow 0$; $x_i \leftarrow a_{in} - \dfrac{1}{2} \displaystyle\sum_{j=1}^{n} a_{ij}$ $(i = 1, n)$; sgn $\leftarrow -1$.

(B) sgn $\leftarrow -$sgn; $P \leftarrow$ sgn; Get next subset of $\{1, 2, \ldots, n-1\}$ from NEXSUB (see Chapter 1); If empty, to (C); If j was deleted, $z \leftarrow -1$; Otherwise, $z \leftarrow 1$; $x_i \leftarrow x_i + za_{ij}$ $(i = 1, n)$.

(C) $P \leftarrow P \cdot x_i$ $(i = 1, n)$; $p \leftarrow p + P$; If more subsets remain, to (B); Permanent $\leftarrow 2(-1)^{n-1}p$; Exit ■

SUBROUTINE SPECIFICATIONS

(1) *Name of subroutine:* PERMAN.
(2) *Calling statement:* CALL PERMAN(N,A,IN,X,PERMN).
(3) *Purpose of subroutine:* Calculate permanent of square matrix.
(4) *Descriptions of variables in calling statement:*

Name	Type	I/O/W/B	Description
N	INTEGER	I	Size of input matrix.
A	DOUBLE PRECISION(N,N)	I	A(I,J) is the I,J entry of the input matrix.
IN	INTEGER(N)	W	Working storage.
X	DOUBLE PRECISION(N)	W	Working storage.
PERMN	DOUBLE PRECISION	O	Calculated value of the permanent of A.

(5) *Other routines which are called by this one:* NEXSUB.
(6) *Number of* FORTRAN *instructions:* 26.
(7) *Remarks:* Will not work correctly in integer arithmetic by merely changing type declarations.

```
      SUBROUTINE PERMAN(N,A,IN,X,PERM)
      IMPLICIT DOUBLE PRECISION(A-H,O-Z)
      LOGICAL MTC
      DIMENSION A(N,N),IN(N),X(N)
10    P=0
      N1=N-1
      DO 11  I=1,N
      SUM=0
      DO 15  J=1,N
15    SUM=SUM+A(I,J)
```

```
11   X(I)=A(I,N)-SUM/2.D0
     SGN=-1
20   SGN=-SGN
     PROD=SGN
30   CALL NEXSUB(N1,IN,MTC,NCARD,J)
     IF(NCARD.EQ.0) GO TO 38
     Z=2*IN(J)-1
     DO 35   I=1,N
35   X(I)=X(I)+Z*A(I,J)
38   DO 39   I=1,N
39   PROD=PROD*X(I)
     P=P+PROD
     IF(MTC) GO TO 20
40   PERM=2.*(2*MOD(N,2)-1)*P
     RETURN
     END
```

SAMPLE OUTPUT

For each $n = 2, 3, \ldots, 12$, PERMAN was asked for the permanent of the $n \times n$ matrix of diagonal 0's and off-diagonal 1's. The output is reproduced below.

2	0.10000000D+01
3	0.20000000D+01
4	0.90000000D+01
5	0.44000000D+02
6	0.26500000D+03
7	0.18540000D+04
8	0.14833000D+05
9	0.13349600D+06
10	0.13349610D+07
11	0.14684570D+08
12	0.17621484D+09

24

Invert a Triangular Array (INVERT)

This little routine is combinatorial only in its proposed use. In fact it is simple linear algebra. We suppose that there is given an $n \times n$ matrix A which has 1's on the main diagonal and 0's below the main diagonal. If b_{ij} $(i, j = 1, n)$ are the entries of A^{-1}, then it is simple to verify that

$$(1) \qquad b_{ij} = \begin{cases} -\sum_{i<k\leq j} a_{ik}b_{kj} & (i < j) \\ 1 & (i = j) \\ 0 & (i > j) \end{cases}$$

We have, then, a simple recurrence formula for calculating the b_{ij} in the following order

$$b_{jj}, b_{j-1,j}, b_{j-2,j}, \ldots, b_{1,j} \quad (j = n, \ldots, 1)$$

The computation is straightforward. It is important to notice that the columns are processed in reverse order to permit storage of A and A^{-1} in the same memory space if desired.

ALGORITHM INVERT

(A) $b_{ii} \leftarrow 1$ $(i = 1, n)$.

(B) $\left(b_{ij} \leftarrow - \sum_{i<k\leq j} a_{ik}b_{kj}, \; (i = j-1, j-2, \ldots, 1), \; j = n, \ldots, 2 \right)$;

  Exit ■

SUBROUTINE SPECIFICATIONS

(1) *Name of subroutine:*   INVERT.
(2) *Calling statement:*   CALL INVERT(A,AINV,N).
(3) *Purpose of subroutine:*   Invert upper triangular matrix.
(4) *Descriptions of variables in calling statement:*

Name	Type	I/O/W/B	Description
A	INTEGER(N,N)	I	Input array.
AINV	INTEGER(N,N)	O	Output, inverse of A.
N	INTEGER	I	Size of A.

(5) *Other routines which are called by this one:*   None.
(6) *Number of* FORTRAN *instructions:*   17.
(7) *Remarks:*   If this subroutine is called with AINV=A, then the matrix A will be correctly inverted *in place*.

```
      SUBROUTINE INVERT (N,A,AINV)
      IMPLICIT INTEGER(A-Z)
      DIMENSION A(N,N),AINV(N,N)
      J=N
10    I=N
20    SUM=0
      IF (I.EQ.J) SUM=1
      K=I+1
25    IF (K.GT.J) GO TO 30
      SUM=SUM-A(I,K)*AINV(K,J)
      K=K+1
      GO TO 25
30    AINV (I,J)=SUM
      I=I-1
      IF (I.GT.0) GO TO 20
      J=J-1
      IF (J.GT.0) GO TO 10
      RETURN
      END
```

25

Triangular Numbering in Partially Ordered Sets (TRIANG)

Let $\mathcal{P}$ be a finite partially ordered set, say

$$\mathcal{P} = \{1, 2, \ldots, n\}$$

and let $\leqslant$ be the partial-order relation defined on $\mathcal{P}$. By the *zeta matrix* of $\mathcal{P}$ we mean the incidence matrix ζ of the relation $\leqslant$ i.e.,

$$(1) \qquad \zeta_{ij} = \begin{cases} 1 & \text{if} \quad i \leqslant j \\ 0 & \text{otherwise} \end{cases}$$

We claim that it is always possible to renumber the rows and columns of ζ (i.e., relabel the elements of $\mathcal{P}$) in such a way that ζ becomes an upper-triangular matrix. In terms of the original partially ordered set $\mathcal{P}$, we are claiming that there is a permutation

$$\sigma: \{1, \ldots, n\} \to \{1, \ldots, n\}$$

such that

$$(2) \qquad \sigma^{-1}(i) \leqslant \sigma^{-1}(j) \Rightarrow i \leq j \quad (i, j = 1, \ldots, n)$$

where the $\leq$ on the right side of (2) is the natural order of the positive integers. Such renumbering will be used in Chapter 26 to construct the Möbius function and is useful in many combinatorial situations where computation must be sequenced consistently with some natural partial order in the problem.

We prove the claim by describing an algorithm which accomplishes the desired renumbering. Choose an element $x \in \mathcal{P}$. If there is no $y \prec x$ ($y \in \mathcal{P}$), then assign to x the next available label. If there is such a y, replace x by y and repeat. Since $\mathcal{P}$ is finite, we surely will reach an element with no predecessor after a finite number of steps.

As is so often the case, the most obvious algorithm is not the best one.‡ In the algorithm described above, we descend a chain in the partial-order relation until we reach a minimal element. Next we search for a new unlabeled element and repeat the process. However, we have lost a good deal of useful information because the element which preceded the minimal element just labeled is a better place to start the search for the next element to label. This is so because we would thereby start lower down in the partially ordered set and would be nearer to our next minimal element.

To recover this information we must make better use of our array σ_i ($i = 1, n$) which carries the labels of the points. This array is set to zero initially, and if a point is labeled, then σ_i carries the label. However, we will now put σ_i to work at intermediate stages also. Precisely, as we go down a chain $i_1 > i_2 > \cdots > i_{\mu-1} > i_\mu$ we write in each σ_{i_k} its predecessor i_{k-1} ($k = 2, \mu$). Then, when we reach i_μ, we save σ_{i_μ} temporarily, in q, say, insert the label of i_μ into σ_{i_μ}, and then resume the search at q.

If q is zero then we have just labeled a point which was the largest element i_1 in some chain of $\mathcal{P}$, all of whose elements are now labeled. Therefore, we resume our search for the next unlabeled point at $1 + i_1$. If q is not zero, then q is indeed the predecessor of the point which was just labeled, and we climb down a chain hanging below q. The nature of the numbering process is such that when we are searching below q to find a minimal element, the search can begin at the next integer larger than the last i below q which was labeled. If we begin the search below q, we may start at $i_1 + 1$ because all points before i_1 have already been labeled.

We give below the formal algorithm which describes the process in detail. This algorithm is designed so that the partial-order relation can be described on input by its full zeta matrix, or else by just the

‡ One considerably more efficient method for this problem, called "topological sorting," appears in [K1, Vol. 1, p. 262]. It is assumed there that the input is given as a set of related ordered pairs. For our purposes we need a square incidence matrix of input.

matrix of its covering relation, or more generally, by any matrix whose (i, j) entry is nonzero if j covers i and 0 if $j \prec i$. In fact, if one is assured that the full zeta matrix will be used for input, then certain further economies become possible: In step **(B)** we can delete "$t \leftarrow m + 1$," and step **(C)** can be replaced by "$r \leftarrow m + 1$" which will start the searches lower down in the structure.

ALGORITHM TRIANG

(A) $m \leftarrow 0$; $l \leftarrow 0$; $\sigma_i \leftarrow 0$ $(i = 1, n)$.

(B)   [*To next unlabeled*] $m \leftarrow m + 1$; If $\sigma_m \neq 0$, to (E); $t \leftarrow m + 1$.

(C)   [*Start climb down*] $r \leftarrow t$.

(D)   If $r \leq n$, to (F); [*Label m*] $l \leftarrow l + 1$; $q \leftarrow \sigma_m$; $\sigma_m \leftarrow l$; [*Climb up to predecessor*] If $q = 0$, to (E); $r \leftarrow m + 1$; $m \leftarrow q$; To (D).

(E)   [*Done?*] If $m = n$, exit; To (B).

(F)   [Is r an unlabeled element below m?] If $\sigma_r \neq 0$ or $\zeta(r, m) \neq 1$, set $r \leftarrow r + 1$ and return to (D); [*Go down from m to r*] $\sigma_r \leftarrow m$; $m \leftarrow r$; To (C) ■

SUBROUTINE SPECIFICATIONS

(1)   *Name of subroutine:*   TRIANG.

(2)   *Calling statement:*   CALL TRIANG(N,ZETA,SIG).

(3)   *Purpose of subroutine:*   Discover consistent labeling of elements of partially ordered set.

(4)   *Descriptions of variables in calling statement:*

Name	Type	I/O/W/B	Description
N	INTEGER	I	Number of elements in partially ordered set.
ZETA	INTEGER(N,N)	I	ZETA(I,J)=1 if I ≤ J, 0 otherwise.
SIG	INTEGER(N)	O	SIG(I) is the new label assigned to I (1≤SIG(I)≤N;1≤I≤N).

(5)   *Other routines which are called by this one:*   None.

(6)   *Number of* FORTRAN *instructions:*   20.

(7)   *Remarks:* Input matrix can be any matrix which generates the partial order.

```
      SUBROUTINE TRIANG(N,ZETA,SIG)
      IMPLICIT INTEGER(A-Z)
      DIMENSION SIG(N), ZETA(N,N)
10    M=0
      L=0
      DO 11 I=1,N
11    SIG(I)=0
20    M=M+1
30    IF (SIG(M).EQ.0) GO TO 40
130   IF (M.EQ.N) RETURN
      GO TO 20
40    T=M+1
50    R=T
60    IF (R.GT.N) GO TO 100
70    IF (SIG(R).NE.0.OR.ZETA(R,M).EQ.0) GO TO 90
80    SIG(R)=M
      M=R
      GO TO 50
90    R=R+1
      GO TO 60
100   L=L+1
      Q=SIG(M)
      SIG(M)=L
110   IF (Q.EQ.0) GO TO 130
      R=M+1
120   M=Q
      GO TO 60
      END
```

SAMPLE OUTPUT

The partially ordered set in our example is the set of divisors of 48 ordered by divisibility. The divisors are arranged, on input, in the order

$$16, \quad 3, \quad 8, \quad 24, \quad 1, \quad 6, \quad 2, \quad 12, \quad 48, \quad 4$$

The zeta matrix corresponding to this input ordering is shown as the following 10×10 array. The output permutation σ, which also follows, rearranges the divisors in the order

$$1, \quad 2, \quad 4, \quad 8, \quad 16, \quad 3, \quad 6, \quad 12, \quad 24, \quad 48$$

in which the zeta matrix is triangular.

```
1  0  0  0  0  0  0  0  1  0
0  1  0  1  0  1  0  1  1  0
1  0  1  1  0  0  0  0  1  0
0  0  0  1  0  0  0  0  1  0
1  1  1  1  1  1  1  1  1  1
0  0  0  1  0  1  0  1  1  0
1  0  1  1  0  1  1  1  1  1
0  0  0  1  0  0  0  1  1  0
0  0  0  0  0  0  0  0  1  0
1  0  1  1  0  0  0  1  1  1

5  6  4  9  1  7  2  8  10  3
```

26

The Möbius Function (MOBIUS)

Let $\mathscr{P}$ be a partially ordered set, and let f be a function defined on $\mathscr{P}$ to the real numbers. Then we can define a new function g on $\mathscr{P}$ by

$$(1) \qquad g(x) = \sum_{y \leqslant x} f(y) \quad (x \in \mathscr{P})$$

As we noted in Chapter 25, if $\mathscr{P}$ contains a 0 element and is locally finite (which we henceforth assume), then the sum in (1) has only a finite number of terms in it for each $x \in \mathscr{P}$. We can rewrite (1) as

$$(2) \qquad g(x) = \sum_{y} \zeta_{x,y} f(y) \quad (x \in \mathscr{P})$$

where the sum is now over all $y \in \mathscr{P}$, and ζ is the zeta function (see Chapter 25) of $\mathscr{P}$. In simple vector-matrix form we could write (2) as

$$(3) \qquad g = \zeta f$$

We now ask how to invert the relation (1). That is, if $g(x)$ is given, for all $x \in \mathscr{P}$, how can we find $f(x)$ $(x \in \mathscr{P})$ such that (1) is true? The importance of this question rests on the fact that in many combinatorial situations where we want f, f and g are related by (1), and g is relatively easy to find.

Now (3) suggests that the inverse relation is

(4) $$f = \zeta^{-1} g$$

provided ζ has an inverse. Our discussion in Chapter 25, however, showed that, under the present hypotheses, the elements of $\mathscr{P}$ can be relabeled so that ζ is upper-triangular with 1's on the diagonal. Such a matrix can always be inverted (see Chapter 24).

The *Möbius function* $\mu(x, y)$ $(x, y \in \mathscr{P})$ is defined as the inverse of the zeta function $\zeta_{x, y}$ $(x, y \in \mathscr{P})$ of $\mathscr{P}$. Our problem in this chapter concerns the efficient computation of $\mu(x, y)$ for a "given" set $\mathscr{P}$.

First we ask for an efficient way to describe the given set $\mathscr{P}$. Certainly the zeta matrix completely describes $\mathscr{P}$. On the other hand, it contains a good deal of redundant information. If we are told, for example, that $\zeta_{1,3} = 1$ and $\zeta_{3,7} = 1$, we do not need to be told that $\zeta_{1,7} = 1$ since that follows from the transitivity of the $\leqslant$ relation.

To describe a more economical method, we define the "covering" relation. We say that, in a partially ordered set $\mathscr{P}$, b *covers* a, written $a \, c \, b$, if

(i) $a \prec b$

and

(ii) there is no $z \in \mathscr{P}$ such that $a \prec z \prec b$.

We denote by $H(x, y)$ the incidence matrix of the covering relation

$$H(x, y) = \begin{cases} 1 & \text{if } x \, c \, y \\ 0 & \text{otherwise} \end{cases}$$

The H matrix describes the complete partial-order relation in $\mathscr{P}$, for if $x \, \alpha \, y$, then there is a chain

$$x \, c \, x_1 \, c \, x_2 \, c \, \cdots \, c \, x_p \, c \, y$$

in which each element is covered by its successor joining x to y. By finding the totality of such chains, we can therefore deduce the totality of relations $x \prec y$, and thereby find the full zeta matrix from its "skeleton" H.

Let us describe this process more formally. Let $\mathscr{P}$ be a finite partially ordered set. Then

$$H^2(x, y) = \sum_z H(x, z) H(z, y)$$

according to the rules of matrix multiplication. A term on the right side is 0 unless z covers x and is covered by y. The sum therefore

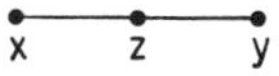

Fig. 26.1

counts the number of maximal chains of length 2 (Fig. 26.1) which join x to y. Similarly, $H^k(x, y)$ counts the number of maximal chains of length k

$$x \text{ c } z_1 \text{ c } z_2 \text{ c } \cdots \text{ c } z_{k-1} \text{ c } y$$

which join x to y, in which each element is covered by its successor.

Now if $x \prec y$ in $\mathscr{P}$, there surely is a chain of *some* length which joins x to y, with each element covered by its successor. Hence, at least one of the numbers

$$H^k(x, y) \quad (k = 1, 2, 3, \ldots)$$

must be positive. Since the others are nonnegative, it follows that

(5) $$H(x, y) + H^2(x, y) + H^3(x, y) + \cdots$$

will have positive entries precisely where $x \prec y$, or, equivalently,

(6) $$\delta(x, y) + H(x, y) + H^2(x, y) + \cdots$$

has positive entries precisely where $x \preccurlyeq y$ (δ is the Kronecker delta).

The apparently infinite series (6) actually terminates. Indeed, since $\mathscr{P}$ is finite, there is only a finite number of different covering chains in $\mathscr{P}$, and if N is the length of the largest one, then $H^m(x, y) = 0$ for all $m > N$ and all $(x, y) \in \mathscr{P}$. The series (6) therefore represents $(I - H)^{-1}$, where I is the identity matrix, and we have shown the

Proposition *In a finite partially ordered set $\mathscr{P}$ with covering matrix H, we have $x \preccurlyeq y$ if and only if*

(7) $$(I - H)^{-1}_{x,y} > 0$$

This relation remains true if H is any nonnegative matrix which generates the partial order.

More precisely, this proposition is true if we have only

(a) $H_{x,y} = 0$ when $x \preccurlyeq y$ is false
(b) $H_{x,y} > 0$ when $x \text{ c } y$
(c) $H_{x,y} \geqq 0$ always

For notational convenience, let us define for any matrix Q of nonnegative entries a new matrix $\psi(Q)$ according to

$$(8) \qquad \psi(Q)_{i,j} = \begin{cases} 1 & \text{if } Q_{i,j} > 0 \\ 0 & \text{if } Q_{i,j} = 0 \end{cases}$$

According to the proposition above, then, the zeta matrix of a finite partially ordered set can be generated from the covering matrix H by means of the relation

$$(9) \qquad \zeta = \psi((I - H)^{-1})$$

Finally, the Möbius function μ is the inverse of ζ, and so it can be obtained from the covering matrix by

$$(10) \qquad \mu = \psi((I - H)^{-1})^{-1}$$

The covering matrix H can be obtained from the zeta matrix in a similar manner because $x \, c \, y$ if and only if there is precisely one chain

$$x = x_0 \prec x_1 \prec \cdots \prec x_p = y$$

In that case, p must be 1. It is easily seen that the (i, j) entry in

$$(\zeta - I) + (\zeta - I)^2 + \cdots$$

counts the number of chains from i to j.

Define for any matrix Q a new matrix $\omega(Q)$ according to

$$\omega(Q)_{i,j} = \begin{cases} 1 & \text{if } i \neq j \text{ and } Q_{i,j} = 1 \\ 0 & \text{otherwise} \end{cases}$$

then we have

$$H = \omega((2I - \zeta)^{-1})$$

We hereby summarize the calculation. Beginning with the incidence matrix H of the a covers b relation, we perform the following operations:

(1) Find the permutation SIGMA such that when the rows and columns of H are renumbered according to SIGMA, H becomes triangular (use subroutine TRIANG, Chapter 25).
(2) Apply the permutation SIGMA to the rows and columns of H (use subroutine RENUMB, Chapter 17).
(3) Invert $I - H$ (use subroutine INVERT, Chapter 24).
(4) Replace all nonzero elements by 1's, yielding the zeta matrix of the partial order.
(5) Invert the resulting matrix (use subroutine INVERT, Chapter 24). This gives the Möbius function MU but according to the renumbering SIGMA.

(6) Find the inverse permutation SIG1 of SIGMA. Renumber the rows and columns of MU according to SIG1 (use subroutine RENUMB, Chapter 17). We now have the Möbius matrix MU consistent with the initial ordering of the rows and columns of H. Output MU. Exit.

Another procedure for doing the above would be to ignore the renumbering and simply invert the matrices as they are. If this were done, we would need to use general matrix inversion programs because the triangularity would no longer be exploited. Thus, instead of INVERT, which requires $\frac{1}{6}n^3 + O(n^2)$ operations to invert a triangular matrix, we would use a program which, at best, might need $\frac{1}{2}n^3$ operations plus the inconvenience of dealing with real numbers, or three times the computational effort. The cost of triangularizing, renumbering, and unrenumbering is $O(n^2)$, which gives an economic advantage to the procedure outlined above.

An additional noteworthy feature is that the entire calculation can be done with just one integer matrix array because each matrix can be written over its predecessor.

SUBROUTINE SPECIFICATIONS

(1) *Name of subroutine:* MOBIUS.

(2) *Calling statement:* CALL MOBIUS(N,H,MU,SIGMA,SIG1).

(3) *Purpose of subroutine:* Find Möbius matrix from covering relation.

(4) *Descriptions of variables in calling statement:*

Name	Type	I/O/W/B	Description
N	INTEGER	I	Number of elements in partially ordered set P.
H	INTEGER(N,N)	I	H(I,J)=1 if I is covered by J, 0 otherwise (I,J=1,N).
MU	INTEGER(N,N)	O	MU(I,J) is Möbius matrix element (I,J=1,N).
SIGMA	INTEGER(N)	W	Working storage.
SIG1	INTEGER(N)	W	Working storage.

(5) *Other routines which are called by this one:* TRIANG, RENUMB, INVERT.

(6) *Number of* FORTRAN *instructions:* 23.

(7) *Remarks:* If called with MU and H being the same array, the cor-

rect output will be obtained, and the input H will of course be lost. Then the routine requires only a single square integer array of storage.

```
SUBROUTINE MOBIUS(N,H,MU,SIGMA,SIG1)
IMPLICIT INTEGER(A-Z)
DIMENSION H(N,N),SIGMA(N),MU(N,N),SIG1(N)
CALL TRIANG(N,H,SIGMA)
DO 1 I=1,N
DO 1 J=1,N
1    MU(I,J)=H(I,J)
CALL RENUMB(N,N,SIGMA,SIGMA,MU)
N1=N-1
DO 11  I=1,N1
J1=I+1
DO 11  J=J1,N
11   MU(I,J)=-MU(I,J)
CALL INVERT(N,MU,MU)
DO 12  I=1,N
DO 12  J=I,N
12   IF(MU(I,J).NE.0)MU(I,J)=1
CALL INVERT(N,MU,MU)
DO 20  I=1,N
20   SIG1(SIGMA(I))=I
CALL RENUMB(N,N,SIG1,SIG1,MU)
RETURN
END
```

SAMPLE OUTPUT

On the following page we show the input H matrix and the output μ matrix in the case of the set of partitions of the integer 6, partially ordered by refinement. The order is: (1) $4 + 1 + 1$, (2) $2 + 1 + 1 + 1 + 1$, (3) $3 + 1 + 1 + 1$, (4) $2 + 2 + 1 + 1$, (5) $2 + 2 + 2$, (6) $4 + 2$, (7) 6, (8) $1 + 1 + 1 + 1 + 1 + 1$, (9) $3 + 2 + 1$, (10) $5 + 1$, (11) $3 + 3$.

For example,

$$\mu(2 + 2 + 1 + 1, 4 + 2) = 2 = \mu(3 + 2 + 1, 6)$$

and all other entries of the μ matrix are at most 1, in absolute value.

```
0  0  0  0  0  1  0  0  0  1  0
0  0  1  1  0  0  0  0  0  0  0
1  0  0  0  0  0  0  0  1  0  0
1  0  0  0  1  0  0  0  1  0  0
0  0  0  0  0  1  0  0  0  0  0
0  0  0  0  0  0  1  0  0  0  0
0  0  0  0  0  0  0  0  0  0  0
0  1  0  0  0  0  0  0  0  0  0
0  0  0  0  0  1  0  0  0  1  1
0  0  0  0  0  0  1  0  0  0  0
0  0  0  0  0  0  1  0  0  0  0
```

```
 1   0   0   0   0  -1   1   0   0  -1   0
 1   1  -1  -1   0  -1   1   0   1  -1   0
-1   0   1   0   0   1  -1   0  -1   1   0
-1   0   0   1  -1   2  -1   0  -1   1   0
 0   0   0   0   1  -1   0   0   0   0   0
 0   0   0   0   0  -1   1   0   0   0   0
 0   0   0   0   0   0   1   0   0   0   0
 0  -1   0   0   0   0   0   1   0   0   0
 0   0   0   0   0  -1   2   0   1  -1  -1
 0   0   0   0   0   0  -1   0   0   1   0
 0   0   0   0   0   0  -1   0   0   0   1
```

27

The Backtrack Method (BACKTR)

(A) GENERAL (BACKTR)

The backtrack method is a reasonable approach to use on problems of exhaustive search when all possibilities must be enumerated or processed. The precise mathematical setting is that we are required to find all vectors

$$(a_1, a_2, \ldots, a_l)$$

of given length l, whose entries $a_1, \ldots, a_l$ satisfy a certain condition $\mathscr{C}$. In the most naive approach, we might first make a list of all possible vectors whose entries a_i lie within the range of the problem; then, from this list we could strike out all vectors which do not satisfy our condition $\mathscr{C}$.

In the backtrack procedure, we "grow" the vector from left to right, and we test at each stage to see if our partially constructed vector has any chance to be extended to a vector which satisfies $\mathscr{C}$. If not, we immediately reject the partial vector, and go to the next one, thereby saving the effort of constructing the descendants of a clearly unsuitable partial vector.

Thus, at the kth stage $(k = 1, l)$, we have before us a partial vector

$$(a_1, a_2, \ldots, a_{k-1})$$

which is not inconsistent with $\mathscr{C}$. We construct from it the list of all candidates for the kth position in our vector. To say that a particular element x is a candidate is just to say that the new partial vector

$$(a_1, a_2, \ldots, a_{k-1}, x)$$

does not yet show any irretrievable inconsistency with our condition $\mathscr{C}$.

If there are no candidates for the kth position, i.e., if for every x, the extended vector $(a_1, a_2, \ldots, a_{k-1}, x)$ is inconsistent with $\mathscr{C}$, we "backtrack" by reducing k by 1, deleting a_{k-1} from the list of candidates for position $k-1$, and choosing the new occupant of the $(k-1)$th position from the reduced list of candidates.

If and when we reach $k = l$, we exit with $a_1, \ldots, a_l$. Upon reentry, we delete a_l from the list of candidates for position l and proceed as before.

We now discuss the computer implementation of this procedure. Our aim is to split off the universal aspects of the backtrack method as a subroutine which will be useful in most, or all, applications, and to leave the part of the application which differs from one situation to the next to the user, as a program which he must prepare within certain guidelines. Our approach is that we suppose that the user wishes to prepare a program which will exhibit one vector at a time which satisfies his condition $\mathscr{C}$, and inform him when no more such vectors exist.

Although it would be simplest to have BACKTR produce one vector (or a negative message) on each call, we do not do so because such a program would have to (a) call the subroutine which provides the list of candidates for each position and hence know the name of this routine, and (b) pass along to this subroutine all the variables, arrays, dimensions, etc., that it needs to operate. These will differ from one application to another.

Instead, the method we have adopted involves the following principles (four examples follow in Sections (B)–(E) of this chapter, which should make the ideas much clearer):

(1) The complete calculation is carried out by three programs: MAIN, BACKTR, CANDTE, of which BACKTR is universal (and appears below) and the other two are prepared by the user.

(2) Communication between the programs is as shown in Fig. 27.1. Note that BACKTR and CANDTE do not speak to each other directly.

(3) MAIN receives input data from the "outside world," and asks BACKTR to inaugurate the search for complete vectors by calling

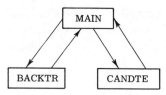

Figure 27.1

BACKTR with INDEX=0.

(4) When BACKTR needs a list of all candidates for the Kth component of the output vector, having already found A(1),...,A(K−1), it asks for this list by RETURN-ing to MAIN with INDEX=2.

(5) MAIN responds to this request by a call to CANDTE, telling CANDTE the value of K, the predecessors A(1),...A(K−1), and whatever auxiliary arrays are needed for the construction.

(6) CANDTE finds the list of candidates and places them at the end of a STACK, i.e., a linear array containing all candidates for all positions up to the Kth, along with a count of the candidates, which becomes the last word in the stack.

(7) MAIN tells BACKTR this information, and the search continues. When K=L, so that the search is successfully completed, BACKTR returns to MAIN with INDEX=1. If there are no more vectors of the type sought, the search terminates with a return to MAIN with INDEX=3.

The above description is a general one. Specific recipes for writing the two routines MAIN and CANDTE will now be given.

The structure of MAIN is as follows:

```
      . . .
      DIMENSION A(100), STACK(1000),...
         Obtain input data


      INDEX=0
1     CALL BACKTR(L,A,INDEX,K,M,STACK,NSTK)
      GO TO (10,20,30), INDEX
10    { Process output vector A(1),...,A(L)
      { but do not change it!
      GO TO 1
20    CALL CANDTE(A,K,M,STACK,...)
      GO TO 1
30    . . .
```

The variables and arrays mentioned play very precise roles:

L is the desired length of a complete output vector.

A is the output vector.

INDEX is explained above.

K is the length of a partially constructed vector: A call to CANDTE is a request for position $A(K)$; K is set by BACKTR.

M is the location of the last item on the stack; it is changed by both BACKTR and CANDTE.

STACK is a linear array, of maximal length NSTK, whose appearance at a typical intermediate stage in the calculation is shown in Fig. 27.2.

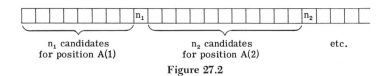

n_1 candidates n_2 candidates   etc.
for position A(1)   for position A(2)

Figure 27.2

More precisely, let the lists of candidates for $A(1),\dots,$ $A(K-1),A(K)$ be stored in STACK, each list followed by its length. Let NC=STACK(M) be the last item on STACK. Then the items STACK(M-1),...,STACK(M-NC) are the candidates for $A(K)$, given the current values of $A(1),\dots,A(K-1)$. When one candidate is needed, NC=STACK(M) is examined; if it is zero, we set M←M-1, K←K-1, and repeat. Otherwise, we set M←M-1, $A(K)$← STACK(M), STACK(M)←NC-1. If K=L, we return A. Otherwise, we set K←K+1 and ask CANDTE to place the candidates for $A(K)$ in locations M+1,...,M+Q of STACK, to enter Q into STACK(M+Q+1), and to set M←M+Q+1. Then BACKTR takes over again.

From this discussion, the precise mission of CANDTE emerges, which we state as follows: *Given* $K,A(1),\dots,A(K-1),M.$ *Find all candidates for* $A(K)$, *insert them in locations* M+1,...,M+Q *of* STACK, *insert* Q *into* STACK(M+Q+1), *set* M←M+Q+1, *and return to* MAIN.

DESCRIPTION OF FLOW CHART

Box 10   Start a new sequence (or first sequence)?

Boxes 20, 30   Initialization to pass information to auxiliary routine.

Box 50   Read length NC of list of candidates for $A(K)$ from stack.

Boxes 60, 70   If NC=0, backtrack to Box 70.

Boxes 80, 90   If K=0, list is complete, exit.

Box 100   Read A(K) from stack, reduce list count.

Boxes 110–130   Vector complete? Set K and INDEX accordingly; exit.

FLOW CHART BACKTR

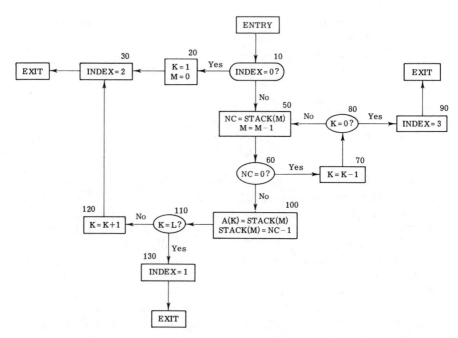

The relationship of backtracking to *random* selection is worth a few remarks. Suppose we have a backtrack situation in which we want a single random choice of an admissible vector, rather than a sequential search for all such vectors. Then, instead of choosing a particular candidate from the list of all possible candidates at the Kth stage, we might envision choosing one of the candidates at random from the list. Unfortunately, if the choice is made uniformly, then not all of the final objects will have equal a priori probability, in general.‡ Thus, this process will, at best, serve as an inadequate substitute for genuine random selection if nothing better suggests itself in a particular situation.

‡ In addition, if there are "few" admissible vectors but many partial vectors, the method may be hard put to find a single admissible vector!

SUBROUTINE SPECIFICATIONS

(1) *Name of subroutine:* BACKTR.
(2) *Calling statement:* CALL BACKTR(L,A,INDEX,K,M,STACK, NSTK).
(3) *Purpose of subroutine:* Supervise backtrack search.
(4) *Descriptions of variables in calling statement:*

Name	Type	I/O/W/B	Description
L	INTEGER	I	Length of completed vector.
A	INTEGER(L)	O	A(1),...,A(L) is an output vector.
INDEX	INTEGER	I/O	=0 to start a search; =1 with a complete output vector; =2 if candidates are needed; =3 if no more vectors exist.
K	INTEGER	I/O	Current length of partial vector.‡
M	INTEGER	I/O	Current length of STACK.‡
STACK	INTEGER(NSTK)	I/O	List‡ of candidates for positions 1,...,K.
NSTK	INTEGER	I	Maximum length of STACK.

‡ Input variables supplied by CANDTE routine.

(5) *Other routines which are called by this one:* None.
(6) *Number of* FORTRAN *instructions:* 23.

```
      SUBROUTINE BACKTR(L,A,INDEX,K,M,STACK,NSTK)
      IMPLICIT INTEGER(A-Z)
      DIMENSION A(L),STACK(NSTK)
10    IF(INDEX.NE.0) GO to 50
20    K=1
      M=0
30    INDEX=2
      RETURN
50    NC=STACK(M)
      M=M-1
60    IF(NC.NE.0) GO TO 100
70    K=K-1
80    IF(K.NE.0) GO TO 50
90    INDEX=3
      RETURN
100   A(K)=STACK(M)
      STACK(M)=NC-1
110   IF(K.NE.L) GO TO 120
      INDEX=1
```

```
      RETURN
120   K=K+1
      GO TO 30
      END
```

(B) COLORING THE VERTICES OF A GRAPH
(COLVRT)

As our first application of backtracking, let G be a graph of n vertices, and let λ be a given positive integer. A proper coloring of the vertices of G in λ colors is an assignment of a color a_i ($1 \leq a_i \leq \lambda$) to each vertex $i = 1, n$ in such a way that for each edge e of G, the two endpoints of e have different colors.

The vector

$$(a_1, a_2, \ldots, a_n)$$

will be the output of our backtrack program, and we will prepare, in this section, the subroutine CANDTE, in this case called COLVRT, which will cause all possible proper colorings of G in λ colors to be delivered sequentially.

We observed in the discussion of Section (A) that the key question for the user of BACKTR is the determination of all candidates for position K of the output vector if a partially constructed vector

$$(A(1), \ldots, A(K-1))$$

is given.

In the present case, our answer is as follows: If K=1, A(1)=1 is the only candidate (for normalization). If K>1, the list of candidates is the set of those integers J such that

(1) $1 \leq J \leq \lambda$

and

(2)   there is no I $\leq$ K−1, such that A(I)=J and vertex I is connected to vertex K in the graph G.

Suppose that the graph G is specified by means of its vertex-adjacency matrix in LOGICAL form, i.e.,

$$ADJ(I,J) = \begin{cases} .TRUE. & \text{if } I < J, \text{ and vertex I connected to J} \\ .FALSE. & \text{if } I < J, \text{ otherwise} \end{cases} \quad (1 \leq I, J \leq N)$$

Then the set of candidates for position K, if K>1, is precisely

$$\{1,2,\ldots,\lambda\}-\{A(I)\,|\,I\leq K-1 \text{ and } ADJ(I,K)=.TRUE.\}$$

The actual program adheres exactly to the format described in Section (A).

SUBROUTINE SPECIFICATIONS

(1) *Name of subroutine:* COLVRT.
(2) *Calling statement:* CALL COLVRT(N,A,K,M,STACK,NSTK, LAMBDA,ADJ,COL).
(3) *Purpose of subroutine:* Find possible colors of vertex K.
(4) *Descriptions of variables in calling statement:*

Name	Type	I/O/W/B	Description
N	INTEGER	I	Number of vertices in the graph.
A	INTEGER(N)	I	A(I) is the color of vertex I(I=1,N).‡
K	INTEGER	I	Vertex whose color-candidates are to be found.‡
M	INTEGER	I/O	Current length of stack.‡
STACK	INTEGER(NSTK)	I/O	Candidates for positions A(1),...,A(K-1).‡
NSTK	INTEGER	I	Maximum length of STACK.
LAMBDA	INTEGER	I	Number of colors available (1≤A(I)≤LAMBDA for I=1,N).
ADJ	LOGICAL(N,N)	I	ADJ(I,J)=.TRUE. if vertices I,J are joined by an edge; =.FALSE. otherwise.
COL	LOGICAL(N)	W	Working storage.

‡ Input values are supplied by BACKTR.

(5) *Other routines which are called by this one:* None.
(6) *Number of* FORTRAN *instructions:* 24.

```
SUBROUTINE COLVRT(N,A,K,M,STACK,NSTK,LAMBDA,ADJ,COL)
IMPLICIT INTEGER(A-Z)
LOGICAL ADJ,COL
DIMENSION A(N),STACK(NSTK),ADJ(N,N),COL(N)
IF(K.GT.1) GO TO 10
STACK(1)=1
STACK(2)=1
M=2
RETURN
```

```
10   K1=K-1
     DO 20   I=1,LAMBDA
20   COL(I)=.TRUE.
     DO 30   I=1,K1
30   IF(ADJ(I,K)) COL(A(I))=.FALSE.
     M1=M
     DO 40   I=1,LAMBDA
     IF(.NOT.COL(I)) GO TO 40
     M1=M1+1
     STACK(M1)=I
40   CONTINUE
     STACK(M1+1)=M1-M
     M=M1+1
     RETURN
     END
```

SAMPLE OUTPUT

The following output shows the six possible proper colorings of a 4-cycle (Fig. 27.3) in which the color of vertex 1 is fixed at color 1.

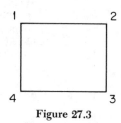

Figure 27.3

The input ADJ array was

$$ADJ= \begin{pmatrix} - & T & F & T \\ T & - & T & F \\ F & T & - & T \\ T & F & T & - \end{pmatrix}$$

```
1   3   2   3
1   3   1   3
1   3   1   2
1   2   3   2
1   2   1   3
1   2   1   2
```

(C) EULER CIRCUITS (EULCRC)

Let G be a (directed or undirected) graph of n vertices and e edges. By an *Euler circuit* on G we mean a walk along the edges of G which visits each edge exactly once, returning to the starting point, and following the direction of each edge if G is directed.

A celebrated theorem of Euler holds that an undirected graph has such a circuit if and only if every vertex of G has even valence. A directed graph has such a circuit if and only if at each vertex there are an equal number of ingoing and outgoing edges. In either case G is called *Eulerian*.

If G is Eulerian we can ask for a program which will list all of the Euler circuits of G in the fashion of our "next" subroutines, i.e., producing one circuit each time called, until no more remain.

The backtrack program provides a ready-made tool for such a task, and so we now describe the utilization of BACKTR in our desired subroutine. As usual, we suppose given a partially constructed Euler circuit

$$(A(1) , A(2) , \ldots , A(K-1))$$

and we ask for the list of candidates for the Kth edge A(K) in the circuit.

We first need to define the idea of a "terminal vertex." If G is undirected, we choose one of the endpoints of edge A(1) and declare it to be the terminal vertex of A(1). Then, for I=2,3,... the terminal vertex Zl(I) is that vertex of A(I) which is not the terminal vertex of A(I-1). If G is directed, the terminal vertex of A(I) is prescribed.

To return now to the question of candidates for A(K), if G is undirected, an edge e' is a candidate for A(K) if

(1) e' does not appear among A(1) , . . . , A(K-1)

and

(2)   the terminal vertex Zl(K-1) of edge A(K-1) is an endpoint of e'.

If G is directed, condition (2) is replaced by

(2')   the terminal vertex Zl(K-1) of edge A(K-1) is the initial vertex of e'.

Euler's theorem guarantees that such a path must return to its starting point, if G is Eulerian.

Many of the interesting applications of our subroutine occur with graphs G which have loops and multiple edges. We therefore permit these in the input graph. Hence input data will consist of ENDPT(1,I),ENDPT(2,I), and the two ends of edge I (I=1,E), where the two endpoints may be equal, and the same pair may appear several times. If G is directed, ENDPT(1,I) is the initial vertex and ENDPT(2,I) is the terminal vertex of edge I.

For an undirected G (OPTION=1), the algorithm for determining the list of candidates for A(K) is

ALGORITHM EULCRC

(A) K=1? If so, set Z1(1)←ENDPT(2,1); Candidate is edge 1, only; Exit.

(B) K=2? If so, to **(D)**.

(C) Z1(K−1)←ENDPT(1,A(K−1))+ENDPT(2,A(K−1))−Z1(K−2).

(D) ED(I)←.FALSE. (I=1,E).

(E) For I=1,E: {If Z1(K−1) is one of the endpoints of edge I, set ED(I)←.TRUE.}.

(F) ED(A(I))←.FALSE. (I=1,K−1) ■

The candidates are then the edges I such that ED(I)=.TRUE.. If G is directed (OPTION=2), step **(E)** above is replaced by

(E′) For I=1,E: {If Z1(K−1) is the initial vertex of edge I, set ED(I)←.TRUE.} ■

A brief comment about step **(C)** above seems warranted. If we have three numbers x, y, z, and if z is known to be one of x, y, but it is not known which one, and if we wish to set w equal to the other one of x, y (i.e., the one which is not z), then the quickest program is

$$w = x + y - z$$

which is done in step **(C)**.

Program CANDTE (called EULCRC) for listing Euler circuits appears on the following two pages.

SUBROUTINE SPECIFICATIONS

(1) *Name of subroutine:* EULCRC.

(2) *Calling statement:* CALL EULCRC(E,A,K,M,STACK,NSTK, OPTION,ENDPT,Z1,ED).

(3) *Purpose of subroutine:* Find candidates for Kth edge of Euler circuit.

(4) *Descriptions of variables in calling statement:*

Name	Type	I/O/W/B	Description
E	INTEGER	I	Number of edges in graph G.
A	INTEGER(E)	I	A(I) is the Ith edge in the circuit (I=1,E).‡
K	INTEGER	I	Index of next edge to be determined in circuit.‡
M	INTEGER	I/O	Current length of stack.‡
STACK	INTEGER(NSTK)	I/O	Candidates for all positions 1,...,K−1.‡
NSTK	INTEGER	I	Maximum length of stack.
OPTION	INTEGER	I	=1 if G is undirected; =2 if G is directed.
ENDPT	INTEGER(2,E)	I	ENDPT(1,I), ENDPT(2,I) are the two ends of edge I(I=1,E).
Z1	INTEGER(E)	W	Working storage.
ED	LOGICAL(E)	W	Working storage.

‡ Input supplied by BACKTR.

(5) *Other routines which are called by this one:* None.

(6) *Number of* FORTRAN *instructions:* 32.

```
      SUBROUTINE EULCRC(E,A,K,M,STACK,NSTK,OPTION,
     *ENDPT,Z1,ED)
      IMPLICIT INTEGER(A-Z)
      LOGICAL ED(E)
      DIMENSION A(E),STACK(NSTK),ENDPT(2,E),Z1(E)
10    IF(K.NE.1) GO TO 30
20    Z1(1)=ENDPT(2,1)
      STACK(1)=1
      STACK(2)=1
      M=2
      RETURN
30    IF(K.EQ.2) GO TO 60
40    Z1(K-1)=ENDPT(1,A(K-1))+ENDPT(2,A(K-1))-Z1(K-2)
60    T=Z1(K-1)
      IF(OPTION.EQ.2) GO TO 80
61    DO 62  I=1,E
62    ED(I)=T.EQ.ENDPT(1,T).OR.T.EQ.ENDPT(2,I)
64    K1=K-1
65    DO 66  I=1,K1
```

```
66   ED(A(I))=.FALSE.
70   M1=M
     DO 71  I=1,E
     IF(.NOT.ED(I)) GO TO 71
     M1=M1+1
     STACK(M1)=I
71   CONTINUE
     STACK(M1+1)=M1-M
     M=M1+1
     RETURN
80   DO 81  I=1,E
81   ED(I)=T.EQ.ENDPT(1,I)
     GO TO 64
     END
```

SAMPLE OUTPUT

The complete graph K_5 on 5 vertices has 132 different Euler circuits. On the following pages there appear, first of all, the ENDPT array which describes K_5, and then the full list of 132 circuits as obtained, successively, from the program.

1	1	1	1	2	2	2	3	3	4
2	3	4	5	3	4	5	4	5	5

1	7	10	8	9	4	3	6	5	2
1	7	10	8	9	4	2	5	6	3
1	7	10	8	5	6	3	4	9	2
1	7	10	8	5	6	3	2	9	4
1	7	10	8	2	4	9	5	6	3
1	7	10	8	2	3	6	5	9	4
1	7	10	6	5	9	4	3	8	2
1	7	10	6	5	9	4	2	8	3
1	7	10	6	5	8	3	4	9	2
1	7	10	6	5	8	3	2	9	4
1	7	10	6	5	2	4	9	8	3
1	7	10	6	5	2	3	8	9	4
1	7	10	3	4	9	8	6	5	2
1	7	10	3	4	9	5	6	8	2
1	7	10	3	2	8	6	5	9	4

1	7	10	3	2	5	6	8	9	4
1	7	9	8	10	4	3	6	5	2
1	7	9	8	10	4	2	5	6	3
1	7	9	8	6	5	2	4	10	3
1	7	9	8	6	5	2	3	10	4
1	7	9	8	3	4	10	6	5	2
1	7	9	8	3	2	5	6	10	4
1	7	9	5	6	10	4	3	8	2
1	7	9	5	6	10	4	2	8	3
1	7	9	5	6	8	2	4	10	3
1	7	9	5	6	8	2	3	10	4
1	7	9	5	6	3	4	10	8	2
1	7	9	5	6	3	2	8	10	4
1	7	9	2	4	10	8	5	6	3
1	7	9	2	4	10	6	5	8	3
1	7	9	2	3	8	5	6	10	4
1	7	9	2	3	6	5	8	10	4
1	7	4	3	10	9	8	6	5	2
1	7	4	3	10	9	5	6	8	2
1	7	4	3	8	9	10	6	5	2
1	7	4	3	8	5	6	10	9	2
1	7	4	3	6	5	9	10	8	2
1	7	4	3	6	5	8	10	9	2
1	7	4	2	9	10	8	5	6	3
1	7	4	2	9	10	6	5	8	3
1	7	4	2	8	10	9	5	6	3
1	7	4	2	8	6	5	9	10	3
1	7	4	2	5	6	10	9	8	3
1	7	4	2	5	6	8	9	10	3
1	6	10	9	8	3	4	7	5	2
1	6	10	9	8	3	2	5	7	4
1	6	10	9	5	7	4	3	8	2
1	6	10	9	5	7	4	2	8	3
1	6	10	9	2	4	7	5	8	3
1	6	10	9	2	3	8	5	7	4
1	6	10	7	5	9	4	3	8	2
1	6	10	7	5	9	4	2	8	3
1	6	10	7	5	8	3	4	9	2
1	6	10	7	5	8	3	2	9	4
1	6	10	7	5	2	4	9	8	3
1	6	10	7	5	2	3	8	9	4
1	6	10	4	3	8	9	7	5	2

1	6	10	4	3	8	5	7	9	2
1	6	10	4	2	9	7	5	8	3
1	6	10	4	2	5	7	9	8	3
1	6	8	9	10	3	4	7	5	2
1	6	8	9	10	3	2	5	7	4
1	6	8	9	7	5	2	4	10	3
1	6	8	9	7	5	2	3	10	4
1	6	8	9	4	3	10	7	5	2
1	6	8	9	4	2	5	7	10	3
1	6	8	5	7	10	3	4	9	2
1	6	8	5	7	10	3	2	9	4
1	6	8	5	7	9	2	4	10	3
1	6	8	5	7	9	2	3	10	4
1	6	8	5	7	4	3	10	9	2
1	6	8	5	7	4	2	9	10	3
1	6	8	2	4	9	5	7	10	3
1	6	8	2	4	7	5	9	10	3
1	6	8	2	3	10	9	5	7	4
1	6	8	2	3	10	7	5	9	4
1	6	3	4	10	8	9	7	5	2
1	6	3	4	10	8	5	7	9	2
1	6	3	4	9	8	10	7	5	2
1	6	3	4	9	5	7	10	8	2
1	6	3	4	7	5	9	10	8	2
1	6	3	4	7	5	8	10	9	2
1	6	3	2	9	10	8	5	7	4
1	6	3	2	9	7	5	8	10	4
1	6	3	2	8	10	9	5	7	4
1	6	3	2	8	10	7	5	9	4
1	6	3	2	5	7	10	8	9	4
1	6	3	2	5	7	9	8	10	4
1	5	9	10	8	2	4	7	6	3
1	5	9	10	8	2	3	6	7	4
1	5	9	10	6	7	4	3	8	2
1	5	9	10	6	7	4	2	8	3
1	5	9	10	3	4	7	6	8	2
1	5	9	10	3	2	8	6	7	4
1	5	9	7	6	10	4	3	8	2
1	5	9	7	6	10	4	2	8	3
1	5	9	7	6	8	2	4	10	3
1	5	9	7	6	8	2	3	10	4
1	5	9	7	6	3	4	10	8	2

1	5	9	7	6	3	2	8	10	4
1	5	9	4	3	10	7	6	8	2
1	5	9	4	3	6	7	10	8	2
1	5	9	4	2	8	10	7	6	3
1	5	9	4	2	8	6	7	10	3
1	5	8	10	9	2	4	7	6	3
1	5	8	10	9	2	3	6	7	4
1	5	8	10	7	6	3	4	9	2
1	5	8	10	7	6	3	2	9	4
1	5	8	10	4	3	6	7	9	2
1	5	8	10	4	2	9	7	6	3
1	5	8	6	7	10	3	4	9	2
1	5	8	6	7	10	3	2	9	4
1	5	8	6	7	9	2	4	10	3
1	5	8	6	7	9	2	3	10	4
1	5	8	6	7	4	3	10	9	2
1	5	8	6	7	4	2	9	10	3
1	5	8	3	4	10	6	7	9	2
1	5	8	3	4	7	6	10	9	2
1	5	8	3	2	9	10	6	7	4
1	5	8	3	2	9	7	6	10	4
1	5	2	4	10	8	9	7	6	3
1	5	2	4	10	6	7	9	8	3
1	5	2	4	9	8	10	7	6	3
1	5	2	4	9	8	6	7	10	3
1	5	2	4	7	6	10	9	8	3
1	5	2	4	7	6	8	9	10	3
1	5	2	3	10	9	8	6	7	4
1	5	2	3	10	7	6	8	9	4
1	5	2	3	8	9	10	6	7	4
1	5	2	3	8	9	7	6	10	4
1	5	2	3	6	7	10	8	9	4
1	5	2	3	6	7	9	8	10	4

(D) HAMILTON CIRCUITS (HAMCRC)

In a graph G of n vertices, a Hamilton circuit is a sequence

$$V_1, V_2, V_3, \ldots, V_n$$

of vertices of G such that the V_i are some rearrangement of all of the vertices of G, each V_i is connected by an edge to V_{i+1} $(i = 1, \ldots, n-1)$, and V_n is connected to V_1.

More pictorially, a Hamilton circuit is a round-trip walk on the edges of G which visits every *vertex* once entering and once leaving (following the directions of the edges if G is a directed graph). Many graphs do not have Hamilton circuits, and there are no simple criteria for deciding whether a given G has such a circuit and, if so, how many different such circuits it has.

The program in this section presents to the calling routine, each time it is called, a Hamilton circuit of G until no more exist, at which time it will so inform the main routine in the usual manner by setting INDEX=3.

We have here a simple exercise in backtracking, in which, if

$$A(1), A(2), \ldots, A(K-1)$$

is the vertex sequence in a partially constructed circuit, the set of candidates for A(K) is the set of all vertices x in G such that

(1) If K=1: $x = 1$.
(2) If K>1: (a) A(K-1) is joined to x by an edge of G
             and
             (b) x is distinct from A(1), ..., A(K-1).
(3) If K=N: (c) x is joined to A(1) by an edge of G
             and
             (d) $x < A(2)$,
             and
             (a) *and* (b) above.

We ensure that each Hamilton circuit occurs exactly once by the normalization conditions in (1) and (3d) above, which require that A(1)=1 and A(N)<A(2).

If G is a directed graph we omit condition (3d). The program will handle an undirected (OPTION=1) or directed (OPTION=2) graph. Program CANDTE (here called HAMCRC) for this purpose appears on the following pages.

SUBROUTINE SPECIFICATIONS

(1) *Name of subroutine:* HAMCRC.

(2) *Calling statement:* CALL HAMCRC(N,A,K,M,STACK,NSTK, ADJ,VERT,OPTION).

(3) *Purpose of subroutine:* Find candidates for Kth vertex in a Hamilton circuit.

(4) *Descriptions of variables in calling statement:*

Name	Type	I/O/W/B	Description
N	INTEGER	I	Number of vertices in graph G.
A	INTEGER(N)	I	A(I) is the Ith vertex in the circuit.‡
K	INTEGER	I	Index of next step on current partial circuit.‡
M	INTEGER	I/O	Current length of stack.‡
STACK	INTEGER(NSTK)	I/O	Candidates for steps 1,...,K−1 (see text).‡
NSTK	INTEGER	I	Maximum length of stack.
ADJ	LOGICAL(N,N)	I	ADJ(I,J)=TRUE if edge (I,J) is in G; FALSE otherwise.
VERT	LOGICAL(N)	W	Working storage.
OPTION	INTEGER	I	=1 if G is undirected; =2 if G is directed.

‡ Input supplied by BACKTR.

(5) *Other routines which are called by this one:* None.

(6) *Number of* FORTRAN *instructions:* 38.

```
      SUBROUTINE HAMCRC(N,A,K,M,STACK,NSTK,ADJ,VERT,OPTION)
      IMPLICIT INTEGER(A-Z)
      LOGICAL ADJ(N,N),VERT(N)
      DIMENSION A(N),STACK(NSTK)
10    IF(K.NE.1) GO TO 30
20    STACK(1)=1
      STACK(2)=1
      M=2
      RETURN
30    K1=K-1
      A1=A(K1)
      DO 31  I=1,N
31    VERT(I)=ADJ(A1,I)
      DO 32  I=1,K1
      M1=A(I)
32    VERT(M1)=.FALSE.
```

```
      M1=M
      IF(K.EQ.N) GO TO 50
40    DO 41  I=1,N
      IF(.NOT.VERT(I)) GO TO 41
      M1=M1+1
      STACK(M1)=I
41    CONTINUE
44    STACK(M1+1)=M1-M
      M=M1+1
      RETURN
50    DO 51  I=1,N
      IF(.NOT.VERT(I)) GO TO 51
      IF(OPTION.EQ.2) GO TO 52
      IF(I.GT.A(2)) GO TO 44
52    IF(.NOT.ADJ(I,1)) GO TO 44
      M=M+2
      STACK(M-1)=I
      STACK(M)=1
      RETURN
51    CONTINUE
      GO TO 44
      END
```

SAMPLE OUTPUT 1

The graph in Fig. 27.4, on 20 vertices, was one of those originally studied by Hamilton. There are 30 Hamilton circuits in the graph,

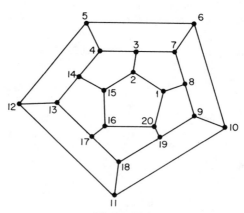

Figure 27.4

and they are shown in the following output, one circuit per line. Elapsed computer time was less than one minute on a relatively slow and small machine.

The vertices of this graph are numbered consecutively following one of the circuits found by Hamilton, which appears as the first one below.

```
20 19 18 17 16 15 14 13 12 11 10  9  8  7  6  5  4  3 2
20 19 18 17 16 15  2  3  7  6  5  4 14 13 12 11 10  9 8
20 19 18 11 12 13 17 16 15 14  4  5  6 10  9  8  7  3 2
20 19 18 11 12  5  6 10  9  8  7  3  4 14 13 17 16 15 2
20 19 18 11 12  5  4 14 13 17 16 15  2  3  7  6 10  9 8
20 19 18 11 10  9  8  7  6  5 12 13 17 16 15 14  4  3 2
20 19  9 10 11 18 17 16 15  2  3  4 14 13 12  5  6  7 8
20 19  9 10  6  5  4 14 13 12 11 18 17 16 15  2  3  7 8
20 19  9  8  7  6 10 11 18 17 16 15 14 13 12  5  4  3 2
20 19  9  8  7  3  4 14 13 12  5  6 10 11 18 17 16 15 2
20 16 17 18 19  9 10 11 12 13 14 15  2  3  4  5  6  7 8
20 16 17 18 19  9  8  7  3  4  5  6 10 11 12 13 14 15 2
20 16 17 13 14 15  2  3  4  5 12 11 18 19  9 10  6  7 8
20 16 17 13 12 11 18 19  9 10  6  5  4 14 15  2  3  7 8
20 16 17 13 12  5  6 10 11 18 19  9  8  7  3  4 14 15 2
20 16 17 13 12  5  4 14 15  2  3  7  6 10 11 18 19  9 8
20 16 15 14 13 17 18 19  9  8  7  6 10 11 12  5  4  3 2
20 16 15 14  4  5  6 10 11 12 13 17 18 19  9  8  7  3 2
20 16 15  2  3  7  6 10 11 12  5  4 14 13 17 18 19  9 8
20 16 15  2  3  4 14 13 17 18 19  9 10 11 12  5  6  7 8
 8  9 19 20 16 17 18 11 10  6  7  3  4  5 12 13 14 15 2
 8  9 19 20 16 15 14  4  5 12 13 17 18 11 10  6  7  3 2
 8  9 10 11 18 19 20 16 17 13 12  5  6  7  3  4 14 15 2
 8  9 10 11 12 13 17 18 19 20 16 15 14  4  5  6  7  3 2
 8  9 10 11 12  5  6  7  3  4 14 13 17 18 19 20 16 15 2
 8  9 10  6  7  3  4  5 12 11 18 19 20 16 17 13 14 15 2
 8  7  6 10  9 19 20 16 15 14 13 17 18 11 12  5  4  3 2
 8  7  6  5 12 13 17 18 11 10  9 19 20 16 15 14  4  3 2
 8  7  3  4 14 13 17 18 11 12  5  6 10  9 19 20 16 15 2
 8  7  3  4  5  6 10  9 19 20 16 17 18 11 12 13 14 15 2
```

SAMPLE OUTPUT 2

Consider the graph G whose 24 vertices correspond to the permutations of 4 letters, where each permutation σ is connected to the three other permutations which can be obtained from σ by a single interchange of two *adjacent* letters. Thus *abcd* is connected to *bacd*, *acbd*, and *abdc*. A Hamilton circuit in G is then a sequencing of the 24 permutations so that each is obtainable from its predecessor by such an interchange.

With the aid of NEXPER, BACKTR, HAMCRC and a main program, we found that there are 44 such Hamilton circuits in G, which are listed on the next page, one to a line, in the following format: To go from permutation I to $I+1$ in the circuit, we exchange the Jth letter with the $(J+1)$th letter ($1 \leq J \leq 3$). The sequence of values of J, for $I=1,23$ is shown in each line.

For example, in the tenth line of the output we find the sequence

$$3212 \quad 1232 \quad 3212 \quad 1232 \quad 3212 \quad 123$$

of values of J, which yields the list of 24 permutations in the following order:

1	1234	9	2314	17	3124
2	1243	10	2341	18	3142
3	1423	11	2431	19	3412
4	4123	12	4231	20	4312
5	4213	13	4321	21	4132
6	2413	14	3421	22	1432
7	2143	15	3241	23	1342
8	2134	16	3214	24	1324

A further examination of this list of Hamilton circuits was carried out in order to find equivalence classes with respect to action of the group generated by (a) replacing each J in a sequence by $3-J$ and (b) cyclically permuting a sequence. This examination showed that there are just five equivalence classes of circuits among the 44 circuits which were printed out, representatives of which are

3121	2131	2121	3123	2121	232
3123	1321	3123	1321	3123	132
3132	3231	3212	3232	1231	323
3212	1232	3212	3212	1232	321
3212	1232	3212	1232	3212	123

```
3 1 2 1 2 1 3 1 2 1 2 1 3 1 2 3 2 1 2 1 2 3 2
3 1 2 1 2 1 3 1 2 3 2 1 2 1 2 3 2 1 3 1 2 1 2
3 1 2 3 2 1 2 1 2 3 2 1 3 1 2 1 2 1 3 1 2 1 2
3 1 2 3 1 3 2 1 3 1 2 3 1 3 2 1 3 1 2 3 1 3 2
3 1 3 2 3 2 3 1 3 2 3 2 3 1 3 2 1 2 3 2 3 2 1
3 1 3 2 3 2 3 1 3 2 1 2 3 2 3 2 1 2 3 1 3 2 3
3 1 3 2 1 3 1 2 3 1 3 2 1 3 1 2 3 1 3 2 1 3 1
3 1 3 2 1 2 3 2 3 2 1 2 3 1 3 2 3 2 3 1 3 2 3
3 2 1 2 1 2 3 2 3 2 1 2 3 2 1 2 1 2 3 2 3 2 1
3 2 1 2 1 2 3 2 3 2 1 2 1 2 3 2 3 2 1 2 1 2 3
3 2 1 2 1 2 3 2 1 3 1 2 1 2 1 3 1 2 1 2 1 3 1
3 2 1 2 1 2 3 2 1 2 3 2 3 2 1 2 1 2 3 2 1 2 3
3 2 1 2 3 1 3 2 3 2 3 1 3 2 3 2 3 1 3 2 1 2 3
3 2 1 2 3 2 3 2 1 2 1 2 3 2 1 2 3 2 3 2 1 2 1
3 2 1 2 3 2 3 2 1 2 3 1 3 2 3 2 3 1 3 2 3 2 3
3 2 1 2 3 2 1 2 1 2 3 2 3 2 1 2 3 2 1 2 1 2 3
3 2 1 3 1 2 1 2 1 3 1 2 1 2 1 3 1 2 3 2 1 2 1
3 2 1 3 1 2 3 1 3 2 1 3 1 2 3 1 3 2 1 3 1 2 3
3 2 3 1 3 2 1 2 3 2 3 2 1 2 3 1 3 2 3 2 3 1 3
3 2 3 1 3 2 3 2 3 1 3 2 1 2 3 2 3 2 1 2 3 1 3
3 2 3 2 1 2 1 2 3 2 1 2 3 2 3 2 1 2 1 2 3 2 1
3 2 3 2 1 2 1 2 3 2 3 2 1 2 1 2 3 2 3 2 1 2 1
3 2 3 2 1 2 3 2 1 2 1 2 3 2 3 2 1 2 3 2 1 2 1
3 2 3 2 1 2 3 1 3 2 3 2 3 1 3 2 3 2 3 1 3 2 1
3 2 3 2 3 1 3 2 1 2 3 2 3 2 1 2 3 1 3 2 3 2 3
3 2 3 2 3 1 3 2 3 2 3 1 3 2 1 2 3 2 3 2 1 2 3
2 3 2 3 2 1 2 1 2 3 2 1 2 3 2 3 2 1 2 1 2 3 2
2 3 2 3 2 1 2 1 2 3 2 3 2 1 2 1 2 3 2 3 2 1 2
2 3 2 3 2 1 2 3 2 1 2 1 2 3 2 3 2 1 2 3 2 1 2
2 3 2 3 2 1 2 3 1 3 2 3 2 3 1 3 2 3 2 3 1 3 2
2 3 2 1 3 1 2 1 2 1 3 1 2 1 2 1 3 1 2 3 2 1 2
2 3 2 1 2 1 2 3 2 3 2 1 2 3 2 1 2 1 2 3 2 3 2
2 3 2 1 2 1 2 3 2 1 3 1 2 1 2 1 3 1 2 1 2 1 3
2 3 2 1 2 3 2 3 2 1 2 1 2 3 2 1 2 3 2 3 2 1 2
2 3 1 3 2 3 2 3 1 3 2 3 2 3 1 3 2 1 2 3 2 3 2
2 3 1 3 2 1 3 1 2 3 1 3 2 1 3 1 2 3 1 3 2 1 3
2 1 3 1 2 3 2 1 2 1 2 3 2 1 3 1 2 1 2 1 3 1 2
2 1 3 1 2 1 2 1 3 1 2 3 2 1 2 1 2 3 2 1 3 1 2
2 1 2 3 2 3 2 1 2 3 2 1 2 1 2 3 2 3 2 1 2 3 2
2 1 2 3 2 3 2 1 2 1 2 3 2 3 2 1 2 1 2 3 2 3 2
2 1 2 3 2 1 3 1 2 1 2 1 3 1 2 1 2 1 3 1 2 3 2
2 1 2 3 2 1 2 3 2 3 2 1 2 1 2 3 2 1 2 3 2 3 2
2 1 2 1 3 1 2 3 2 1 2 1 2 3 2 1 3 1 2 1 2 1 3
2 1 2 1 3 1 2 1 2 1 3 1 2 3 2 1 2 1 2 3 2 1 3
```

(E) SPANNING TREES (SPNTRE)

Our final example of a backtrack routine will, each time it is called, exhibit one spanning tree of a given graph G and inform the user when no more exist.

Suppose, then, that

$$A(1),A(2),\ldots,A(K-1)$$

are the edges of a partially constructed spanning tree of G. What are the candidates for $A(K)$? Suppose we were to adopt the condition that for $A(K)$ we use any edge I such that exactly one endpoint of I is incident with the subgraph spanned by $A(1),\ldots,A(K-1)$. We would surely generate all spanning trees, but a given tree T could be generated many times. This is because the present problem is fundamentally different from the preceding applications in that the order of the components in the output vector is immaterial. A given tree T might appear with several different edge orderings, and we want each one to be generated just once.

To avoid this problem, we might require not only that edge I have exactly one vertex in the previous partial tree, but also that $I>A(K-1)$. This would insure that the edges would be in ascending order, so each tree T would then be generated no more than once. Unfortunately, some trees would not be generated at all! Indeed T could be so generated if and only if when the edges of T are arranged in increasing order of their numbers, each edge $A(I)$ is incident with the subgraph spanned by $A(1),\ldots,A(I-1)$, and this is clearly a special property not shared by all T.

In order to insure that each tree appears at most once, we continue the requirement that $A(K)>A(K-1)$. To insure that all trees appear, we ask only that edge $A(K)$ not form any circuits in the subgraph spanned by $A(1),\ldots,A(K-1)$, but we do not insist that it be incident with that subgraph. Our partially constructed trees $\{A(1),\ldots,A(K-1)\}=T_K$ will actually be forests, i.e., will have several connected components, each of them a tree. To determine whether or not an edge e completes a circuit in T_K we ask if both endpoints of e lie in the same connected component of T_K. Finally, observe that the complete tree T will have $N-1$ edges, and if these edges are to be numbered in ascending order, then we must have $A(K)\leqq E-N+K+1$ $(K=1,\ldots,N-1)$.

To summarize, edge I is a candidate for position K, given

A(1),...,A(K-1) *if*

(1) A(K-1)+1≦I≦E-N+K+1

and

(2) ENDPT(1,I) *and* ENDPT(2,I) *are in different connected components of the subgraph whose vertices are all of the vertices of G and whose edges are* A(1),...,A(K-1).

Subroutine SPANFO of Chapter 18 will assume the task of determining the connected component X(J) in which each vertex J=1,N lives.

SUBROUTINE SPECIFICATIONS

(1) *Name of subroutine:* SPNTRE.

(2) *Calling statement:*
CALL SPNTRE(E,N,A,K,M,STACK,NSTK,ENDPT,END,X)

(3) *Purpose of subroutine:* Find candidates for Kth edge of spanning tree.

(4) *Descriptions of variables in calling statement:*

Name	Type	I/O/W/B	Description
E	INTEGER	I	Number of edges of graph G.
N	INTEGER	I	(Number of vertices of graph G) − 1 (!!!)
A	INTEGER(N)	I	A(I) is Ith edge of spanning tree (I=1,N).‡
K	INTEGER	I	Index of position for which candidates are needed.‡
M	INTEGER	I/O	Current size of stack.‡
STACK	INTEGER(NSTK)	I/O	List of candidates for all positions (see text).‡
NSTK	INTEGER	I	Maximum length of stack.
ENDPT	INTEGER(2,E)	I	ENDPT(1,I), ENDPT(2,I) are the two ends of vertex *I* in G (I=1,E).
END	INTEGER(2,N)	W	Working storage.
X	INTEGER(N)	W	Working storage.

‡ Input supplied by BACKTR.

(5) *Other routines which are called by this one:* SPANFO, RENUMB.

(6) *Number of* FORTRAN *instructions:* 28.

SUBROUTINE SPNTRE(E,N,A,K,M,STACK,NSTK,ENDPT,
*END,X)

```
      IMPLICIT INTEGER(A-Z)
      DIMENSION A(N),STACK(NSTK),ENDPT(2,E),END(2,N),
     *X(N)
10    IF(K.NE.1) GO TO 30
20    N2=E-N+1
      DO 21   I=1,N2
21    STACK(I)=I
      M=N2+1
      STACK(M)=N2
      RETURN
30    K1=K-1
      DO 31   I=1,K1
      END(1,I)=ENDPT(1,A(I))
31    END(2,I)=ENDPT(2,A(I))
      N3=N+1
      CALL SPANFO(N3,K1,END,COMP,X)
      I1=A(K1)+1
      I2=E-N+K
      M1=M
32    DO 35   I=I1,I2
      IF(X(ENDPT(1,I)).EQ.X(ENDPT(2,I))) GO TO 35
      M1=M1+1
      STACK(M1)=I
35    CONTINUE
      STACK(M1+1)=M1-M
      M=M1+1
      RETURN
      END
```

SAMPLE OUTPUT

Suppose that five cities A, B, C, D, E are situated as shown in the "map" in Fig. 27.5 in which the numbers are the distances between cities. We ask for the shortest length of telephone cable which would connect all of the cities together. Evidently, we seek the spanning tree of the graph of shortest total length (Fig. 27.6). The output of our program shows for each of the 125 spanning trees of the graph, first its total length, and then the four edges which comprise the tree. It is seen that the shortest connection is 235 miles long. There are much more efficient ways of handling this problem (see Chapter 30) and this example is intended only to illustrate the operation of SPNTRE.

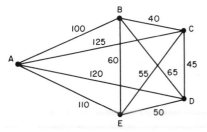

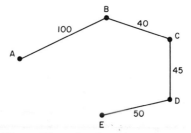

Figure 27.5 The map.   **Figure 27.6** The shortest connection.

455	4	7	9	10	290	3	5	6	7
395	4	7	8	10	400	3	4	7	9
385	4	7	8	9	340	3	4	7	8
375	4	6	9	10	320	3	4	6	9
315	4	6	8	10	260	3	4	6	8
305	4	6	8	9	330	3	4	5	9
380	4	6	7	9	270	3	4	5	8
320	4	6	7	8	345	3	4	5	7
385	4	5	9	10	265	3	4	5	6
325	4	5	8	10	415	2	7	9	10
315	4	5	8	9	355	2	7	8	10
400	4	5	7	10	345	2	7	8	9
330	4	5	7	8	335	2	6	9	10
320	4	5	6	10	275	2	6	8	10
310	4	5	6	9	265	2	6	8	9
325	4	5	6	7	340	2	6	7	9
420	3	7	9	10	280	2	6	7	8
360	3	7	8	10	345	2	5	9	10
350	3	7	8	9	285	2	5	8	10
340	3	6	9	10	275	2	5	8	9
280	3	6	8	10	360	2	5	7	10
270	3	6	8	9	290	2	5	7	8
345	3	6	7	9	280	2	5	6	10
285	3	6	7	8	270	2	5	6	9
350	3	5	9	10	285	2	5	6	7
290	3	5	8	10	405	2	4	7	10
280	3	5	8	9	335	2	4	7	8
365	3	5	7	10	325	2	4	6	10
295	3	5	7	8	255	2	4	6	8
285	3	5	6	10	330	2	4	6	7
275	3	5	6	9	335	2	4	5	10

265	2	4	5	8		295	1	4	6	9
260	2	4	5	6		235	1	4	6	8
370	2	3	7	10		315	1	4	5	10
360	2	3	7	9		245	1	4	5	8
290	2	3	6	10		240	1	4	5	6
280	2	3	6	9		335	1	3	9	10
295	2	3	6	7		275	1	3	8	10
300	2	3	5	10		265	1	3	8	9
290	2	3	5	9		340	1	3	7	9
305	2	3	5	7		280	1	3	7	8
350	2	3	4	7		280	1	3	5	10
270	2	3	4	6		270	1	3	5	9
280	2	3	4	5		285	1	3	5	7
395	1	7	9	10		315	1	3	4	9
335	1	7	8	10		255	1	3	4	8
325	1	7	8	9		260	1	3	4	5
315	1	6	9	10		330	1	2	9	10
255	1	6	8	10		270	1	2	8	10
245	1	6	8	9		260	1	2	8	9
320	1	6	7	9		345	1	2	7	10
260	1	6	7	8		275	1	2	7	8
325	1	5	9	10		265	1	2	6	10
265	1	5	8	10		255	1	2	6	9
255	1	5	8	9		270	1	2	6	7
340	1	5	7	10		320	1	2	4	10
270	1	5	7	8		250	1	2	4	8
260	1	5	6	10		245	1	2	4	6
250	1	5	6	9		285	1	2	3	10
265	1	5	6	7		275	1	2	3	9
370	1	4	9	10		290	1	2	3	7
310	1	4	8	10		265	1	2	3	4
300	1	4	8	9						

28

Labeled Trees (LBLTRE)

We now consider the processing of labeled trees. A celebrated theorem of Cayley asserts that there are exactly n^{n-2} such trees on n vertices. Our plan here is to use one of the proofs of Cayley's theorem as the basis for selection algorithms.

The proof which we will use is due to Prüfer, and it gives an explicit construction which associates with each labeled tree on n vertices a unique $(n-2)$-tuple of integers $a_1, a_2, \ldots, a_{n-2}$ in the range

$$(1) \qquad\qquad 1 \leqq a_i \leqq n \quad (i = 1, n-2)$$

in a 1-1 way. Since there are obviously n^{n-2} sequences ("codewords") $a_1, \ldots, a_{n-2}$ which satisfy (1), Cayley's theorem will follow at once.

Having established this, we will give an algorithm which constructs a tree from its codeword, in linear time. Hence, with this algorithm, one can exhibit all trees on n vertices, select one uniformly at random, find the mth one on the list, etc., since such operations are trivially easy on the codewords themselves.

Given a tree T on n vertices. By an *endpoint* of T we mean a vertex of valence 1. It is easy to see that every tree has at least one endpoint.

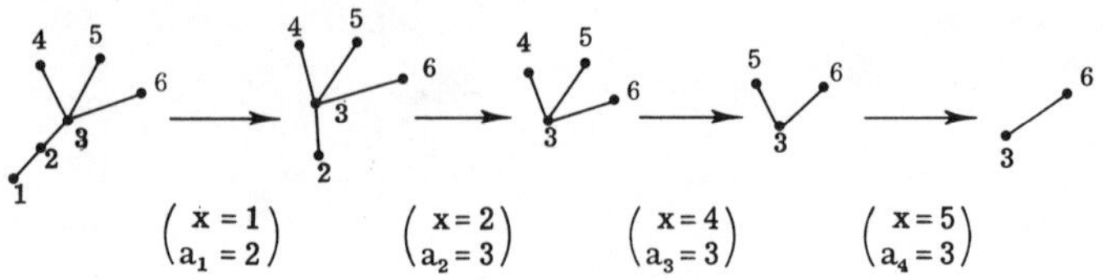

$$\begin{pmatrix} x = 1 \\ a_1 = 2 \end{pmatrix} \quad \begin{pmatrix} x = 2 \\ a_2 = 3 \end{pmatrix} \quad \begin{pmatrix} x = 4 \\ a_3 = 3 \end{pmatrix} \quad \begin{pmatrix} x = 5 \\ a_4 = 3 \end{pmatrix}$$

Figure 28.1

Let x be the endpoint of smallest index. Let a_1 be the unique vertex of T to which x is connected. Delete x and the edge (x, a_1) from T, to obtain a new tree T'. Again, let x' be the endpoint of T' of smallest index, and let a_2 be the unique vertex of T' to which x' is connected. Delete x' and edge (x', a_2) from T' to obtain T'', etc. The process halts when we have found $a_1, a_2, \ldots, a_{n-2}$ and the tree has been reduced to a single edge.

For example, we have the sequence of Fig. 28.1. The tree at the left is associated with the sequence $(2, 3, 3, 3)$ of integers in the range $[1, 6]$.

Prüfer's construction goes both ways. Given a sequence $(a_1, \ldots, a_{n-2})$ in the range $[1, n]$, make two lists. List_1 initially contains the numbers $1, 2, \ldots, n$ in order. List_2 initially contains $a_1, \ldots, a_{n-2}$. List_1 has length 2 greater than List_2. Hence, there are numbers in List_1 which are not in List_2. Let x be the smallest of these $(1 \leqq x \leqq n)$. Connect vertex x and vertex a_1 by an edge. Delete x from List_1 and a_1 from List_2. Again, let x be the smallest number in List_1, which is not in List_2. Connect (x, a_2) by an edge. Delete x and a_2 from their respective lists, etc.

The process terminates when List_2 is empty and List_1 contains two elements x, y. Connect (x, y) by an edge, and the tree is now complete. For example, if $n = 6$, we have the sequence shown in Fig. 28.2.

The argument actually is useful for a good deal more than a proof of Cayley's theorem. For instance, given an n-tuple $\mathbf{a} = (a_1, \ldots,$

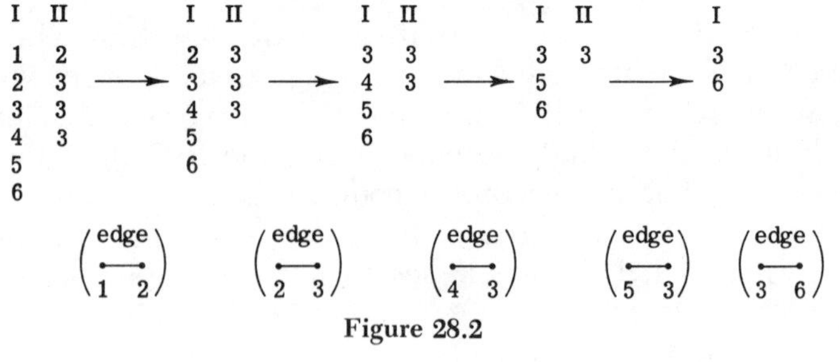

I	II		I	II		I	II		I	II		I
1	2		2	3		3	3		3	3		3
2	3		3	3		4	3		5			6
3	3		4	3		5			6			
4	3		5			6						
5			6									
6												

$$\begin{pmatrix} \text{edge} \\ \bullet\!\!-\!\!\bullet \\ 1 \quad 2 \end{pmatrix} \quad \begin{pmatrix} \text{edge} \\ \bullet\!\!-\!\!\bullet \\ 2 \quad 3 \end{pmatrix} \quad \begin{pmatrix} \text{edge} \\ \bullet\!\!-\!\!\bullet \\ 4 \quad 3 \end{pmatrix} \quad \begin{pmatrix} \text{edge} \\ \bullet\!\!-\!\!\bullet \\ 5 \quad 3 \end{pmatrix} \quad \begin{pmatrix} \text{edge} \\ \bullet\!\!-\!\!\bullet \\ 3 \quad 6 \end{pmatrix}$$

Figure 28.2

a_{n-2}) satisfying (1); what is the valence $\rho(i)$ of vertex i in the tree which corresponds to **a**? The construction shows that $\rho(i) = 1 + \mu(i)$ where $\mu(i)$ is the number of appearances of i in **a** $(i = 1, \ldots, n)$.

Hence, if the number of trees on n labeled vertices with valences $\rho(1), \ldots, \rho(n)$ (given) is denoted by $F_n(\boldsymbol{\rho})$, we know that $F_n(\boldsymbol{\rho})$ is the number of $(n-2)$-tuples **a** which satisfy both (1) and the additional condition that

(2) $\qquad\qquad \mu(i) = \rho(i) - 1 \quad (i = 1, \ldots, n)$

Since evidently

(3) $$\sum_{i=1}^{n} \mu(i) = n - 2$$

we have from (2)

(4) $$\sum_{i=1}^{n} \rho(i) = 2n - 2$$

as a necessary condition on the $\boldsymbol{\rho}$ if $F_n(\boldsymbol{\rho}) > 0$. If (3) holds, then $F_n(\boldsymbol{\rho})$ is the number of ways of arranging $\rho(1) - 1$ 1's, $\rho(2) - 1$ 2's, $\ldots$, $\rho(n) - 1$ n's in an $(n-2)$-vector; i.e.,

(5) $$F_n(\boldsymbol{\rho}) = \frac{(n-2)!}{(\rho(1) - 1)! \cdots (\rho(n) - 1)!}$$

is the number of trees with valence vector $\boldsymbol{\rho}$. Summation of (5) over all $\boldsymbol{\rho}$ which satisfy (4) yields Cayley's theorem again, but (5) is considerably more precise.

How many labeled trees on n vertices have exactly t endpoints? We can select which t vertices shall be the endpoints in

$$\binom{n}{t}$$

ways. The number of trees in which vertices $1, 2, \ldots, t$ are the endpoints is the number of ways of placing $n - 2$ labeled balls into $n - t$ labeled boxes with no box empty. To see this, let the balls be labeled $1, 2, \ldots, n - 2$, and let the boxes be labeled $t + 1, t + 2, \ldots, n$. For any arrangement of the balls in the boxes with no box empty, interpret the set of labels on the balls in box i as the set of subscripts j such that $a_j = i$ $(j = 1, \ldots, n - 2;$ $i = t + 1, \ldots, n)$. The arrangement therefore leads uniquely to a vector $(a_1, \ldots, a_{n-2})$ in which $a_i \geq t + 1$ $(i = 1, \ldots, n - 2)$, and therefore to a tree in which vertices $1, 2, \ldots, t$ are endpoints. We have proved the

Theorem The number of labeled trees on n vertices which have exactly t endpoints is

(6) $$\frac{n!}{t!} \begin{Bmatrix} n-2 \\ n-t \end{Bmatrix} \quad (2 \leqq t \leqq n-1)$$

in which the quantity in braces is a Stirling number of the second kind.

A consequence of (6) is the identity

(7) $$\sum_{t=2}^{n-1} \frac{n!}{t!} \begin{Bmatrix} n-2 \\ n-t \end{Bmatrix} = n^{n-2}$$

which is well-known in the theory of Stirling numbers. The average number of endpoints over all trees on n vertices is

$$\bar{t} = n^{-(n-2)} \sum_{t=2}^{n-1} t \, \frac{n!}{t!} \begin{Bmatrix} n-2 \\ n-t \end{Bmatrix} = n^{-(n-2)} \sum_{r=1}^{n-2} (n-r) \, \frac{n!}{(n-r)!} \begin{Bmatrix} n-2 \\ r \end{Bmatrix}$$

$$= n^{-(n-2)} \cdot n \sum_{r=1}^{n-2} \binom{n-1}{r} r! \begin{Bmatrix} n-2 \\ r \end{Bmatrix} = n^{-(n-2)} \cdot n \cdot (n-1)^{n-2}$$

$$= n \left(1 - \frac{1}{n} \right)^{n-2} \sim \frac{n}{e} \quad (n \to \infty)$$

Hence an average tree has about n/e endpoints.

We return now to the main purpose of the discussion, which is to describe an algorithm for generating a random tree. Evidently what we need to do is just

(a) Select $n-2$ integers $a_1, \ldots, a_{n-2}$ at random in $[1, n]$.
(b) Carry out Prüfer's construction to get the tree.

The formal algorithm utilizes arrays as follows:

A(J)	(J=1,N−2)	The $n-2$ numbers $a_1, \ldots, a_{n-2}$ (this is List₂).
B(J)	(J=1,N)	=.TRUE. if J is on List₁; .FALSE. otherwise.
M(J)	(J=1,N)	The number of appearances of J in List₂.
⎰END(1,M1) ⎱END(2,M1)	(M1=1,N−1)	The two endpoints of the M1th edge in the output tree.

We turn now to the question of implementing Prüfer's construction by a formal algorithm. This algorithm utilizes two linear arrays: a_i

$(i = 1, n - 2)$ is the given codeword, and tree(i) $(i = 1, n - 1)$ describes the output labeled tree T in the sense that $(i, \text{tree}(i))$ $(i = 1, n - 1)$ is the set of edges of T. The algorithm is due to P. Klingsberg.

We begin by flagging the last appearance of each integer m in the codeword by changing its sign. For example, if the input codeword is $(7, 2, 5, 2, 1, 5)$ we would change it to $(-7, 2, 5, -2, -1, -5)$.

Now the array *tree* will hold three kinds of information during the operation of the algorithm:

(a) $Tree(i) = -1$ means that i is still in List_1 and that i still appears in List_2.

(b) $Tree(i) = 0$ means that i is still in List_1 and that i does not appear in List_2 (i is "eligible").

(c) $1 \leq Tree(i) \leq n$ means that i is no longer in List_1 and so Tree(i) is the other endpoint of vertex i in the output tree.

In the example above, the array *tree* would be initially set to $(0, -1, 0, 0, -1, 0, -1, 0)$.

Further, two pointers k, k' are maintained. k' always points to the smallest eligible index i, that is, the least i for which tree(i) $= 0$. The pointer k has the property that for every eligible index i we have $i > k$ or $i = k'$. The formal algorithm follows.

ALGORITHM LBLTRE

[Enter with $a_1, \ldots, a_{n-2}$ ($\forall i\colon 1 \leq a_i \leq n$); Exit with the tree which is thereby encoded, in the form of its edge list $(i, \text{tree}(i))$ $(i = 1, n - 1)$].

(A) [*Initialize*]
　　　tree(i) $\leftarrow 0$ ($i = 1, n$); $k \leftarrow 1$; $j \leftarrow 0$; $a_{n-1} \leftarrow n$.
　　　For $i = n - 2, \ldots, 1$ *do:*
　　　　If tree $(a_i) = -1$, next i;
　　　　tree(a_i) $\leftarrow -1$; $a_i \leftarrow -a_i$; next i.
　　　End
(B) [*Move down to next eligible vertex*]
　　　$k' \leftarrow k \leftarrow \min\{l \geq k | \text{tree}(l) = 0\}$.
(C) [*Enter next edge*] $j \leftarrow j + 1$; $r \leftarrow |a_j|$; tree(k') $\leftarrow r$; [done?] if $j = n - 1$, exit; [last appearance of r on List_2?] If $a_j > 0$, to (B); [reenter r as active letter] if $r > k$ set tree(r) $\leftarrow 0$ and go to (B) $k' \leftarrow r$; to (C) ∎

SUBROUTINE SPECIFICATIONS

(1) *Name of subroutine:* LBLTRE.

(2) *Calling statement:* CALL LBLTRE(N,A,TREE).

(3) *Purpose of subroutine:* Produce the edge list of a tree from its Prüfer codeword.

(4) *Descriptions of variables in calling statement:*

Name	Type	I/O/W/B	Description
N	INTEGER	I	Number of vertices in desired tree.
A	INTEGER(N)	I	A(I)(I=1,N-2) is the Prüfer codeword of the tree.
TREE	INTEGER(N)	O	(I,TREE(I)) is the Ith edge of the output tree (I=1,N-1).

(5) *Other routines which are called by this one:* None.

(6) *Number of* FORTRAN *instructions:* 30.

```
         SUBROUTINE LBLTRE(N,A,TREE)
         INTEGER A(N),TREE(N),R
10       DO 11  I=1,N
11       TREE(I)=0
         NM2=N-2
         DO 12  I=1,NM2
         L=A(N-1-I)
         IF(TREE(L).EQ.0) A(N-1-I)=-L
12       TREE(L)=-1
         K=1
         A(N-1)=N
         J=0
20       IF(TREE(K).EQ.0) GO TO 25
         K=K+1
         GO TO 20
25       KP=K
30       J=J+1
         R=IABS(A(J))
         TREE(KP)=R
         IF(J.EQ.N-1) GO TO 40
32       IF(A(J).GT.0) GO TO 20
         IF(R.GT.K) GO TO 35
         KP=R
```

```
      GO TO 30
35    TREE(R)=0
      GO TO 20
40    DO 41  I=1,NM2
41    A(I)=IABS(A(I))
      RETURN
      END
```

SAMPLE OUTPUT

For our sample problem we used LBLTRE to display the 16 labelled trees on N=4 vertices. On the 16 printed lines below we show the Prüfer codeword A(1),A(2) and the output TREE array, TREE(1),TREE(2),TREE(3), for each of these 16 trees.

```
1  1    4  1  1
2  1    4  1  2
3  1    4  3  1
4  1    4  4  1
1  2    2  4  1
2  2    2  4  2
3  2    3  4  2
4  2    4  4  2
1  3    3  1  4
2  3    2  3  4
3  3    3  3  4
4  3    4  3  4
1  4    4  1  4
2  4    2  4  4
3  4    3  4  4
4  4    4  4  4
```

29

Random Unlabeled Rooted Trees (RANRUT)

The algorithms for finding random partitions of an integer (Ranpar, Chapter 10) and random equivalence classes on a set (Ranequ, Chapter 12) were both based on recurrence relations of the approximate form

(1)
$$na_n = \sum_{m<n} c_{n-m}a_m$$

where a_n is the number of objects of order n and where the c's were known or easy to compute. A combinatorial proof of (1) was essential, which then gave rise to an inductive construction by dividing both sides of (1) by the left side and interpreting the terms on the right as a sum of probabilities that add up to one.

The situation for random unlabeled rooted trees (briefly called "trees" for the remainder of this chapter) is similar since it is also a special case of the ideas in the Postscript to Chapter 10, pp. 78–87, although more complicated. The essential formula which t_n, the number of trees on n vertices, satisfies is now

(2)
$$(n-1)t_n = \sum_{1 \le m < n} t_{n-m} \sum_{d \mid m} d t_d \quad (n > 1, \, t_1 = 1)$$

For n fixed, select an integer m, $1 \le m < n$, a divisor d of m, a tree T' of $n - m$ vertices, and a tree T'' of d vertices. Make $j = m/d$

copies of T''. Join the root R of T' to the roots of each of the copies of T''. There results a tree T of n vertices rooted at R. This operation is symbolized by $T \leftarrow T' + j \otimes T''$.

To prove (2), take d copies of T. We claim that, thus, every rooted tree on n vertices is created exactly $n - 1$ times. Indeed, if T is such a tree, let k be the valence of the root R of T. Delete these k edges and also R, and root each component at the vertex which was connected to R. Suppose the resulting k trees consist of μ_1 copies of a tree τ_1 on l_1 vertices, . . . , μ_s copies of a tree τ_s, on l_s vertices, where, of course, $\mu_1 l_1 + \cdots + \mu_s l_s = n - 1$, and the trees $\tau_1, \ldots, \tau_s$ are nonisomorphic. From these data the tree T can be obtained in the following way: let $1 \le j \le s$, $1 \le r \le \mu_j$, and let T' be the tree obtained by deleting from T r copies of τ_j, including the edges joining their roots to R. Then T is obtained as $T' + r \otimes \tau_j$. We count this construction l_j times; then T is counted $\mu_1 l_1 + \cdots + \mu_s l_s = n - 1$ times, as claimed in (2). See the Postscript to Chapter 10, pp. 78–87, where the general principles underlying such constructions are described.

Formula (2) is well known. The usual proof, which we do not present here, is based on the generating function $T(x) = \Sigma_{n=1}^{\infty} t_n x^n$ which satisfies the identity

$$(3) \qquad T(x) = x \exp\left\{\sum_{r=1}^{\infty} \frac{T(x^r)}{r}\right\}$$

Indeed, our construction proves (3), because (2) and (3) are equivalent after logarithmic differentiation of (3).

Most important for our present purposes, we can now use (2) to construct random trees. First the numbers $t_1, t_2, \ldots, t_n$ are to be computed using (2), or the equivalent form (set $m = jd$; take $t_k = 0$ when $k \le 0$)

$$(4) \qquad (n - 1)t_n = \sum_{d=1}^{\infty} \sum_{j=1}^{\infty} d t_{n-jd} t_d, \quad t_1 = 1$$

then divide by the left side

$$1 = \sum_{j=1}^{\infty} \sum_{d=1}^{\infty} \frac{d t_{n-jd} t_d}{(n - 1)t_n}$$

and interpret the right side as a sum of probabilities: Choose a pair (j, d), with $j \ge 1$, $d \ge 1$, with a priori probability

$$(5) \qquad \text{prob}(j, d) = \frac{d t_{n-jd} t_d}{(n - 1)t_n}$$

then choose (inductively) a random tree T' on $n - jd$ vertices (with probability $\dfrac{1}{t_{n-jd}}$) and a random tree T'' on d vertices (with probability $\dfrac{1}{t_d}$). Carry out the construction described above to yield a rooted tree T. To calculate the a priori probability of T, observe that the present single construction yields T with probability

$$\frac{dt_{n-jd}t_d}{(n-1)t_n} \cdot \frac{1}{t_{n-jd}} \cdot \frac{1}{t_d} = \frac{d}{(n-1)t_n}$$

Suppose now, as before, that T can be constructed from μ_1 copies of a rooted tree $\tau_1, \ldots, \mu_s$ copies of τ_s; then T could have been constructed with τ_1 taking the part of T'' for $j = 1, \ldots, \mu_1$ (so $d = l_1$) with τ_s taking the part of T'' for $j = 1, \ldots, \mu_s$ (so $d = l_s$). The total a priori probability of T is therefore

$$\sum_{m=1}^{s} \sum_{j=1}^{\mu_m} \frac{l_m}{(n-1)t_n} = \frac{1}{(n-1)t_n} \sum_{m=1}^{s} \mu_m l_m = \frac{1}{t_n}$$

and it follows that all trees of n vertices are equally likely to occur.

The actual inductive construction of random rooted unlabeled trees is complicated by the nonlinearity of (2) and (4): the construction of T requires the construction of *two* other random trees, T' and T''; they in their turn require two trees, etc. This does not end until we hit a random tree with 1 or 2 vertices, which we know how to construct.

As an illustration, let $n = 11$. A random selection based on Eq. (5) with $n = 11$ produces, e.g., $(j, d) = (2, 3)$; so $n - jd = 5$. We draw again, now for $n = 5$ (yielding, e.g., $(3, 1)$) and $n = 2$ (yielding $(1, 2)$). We represent the results in a diagram (actually, a binary tree) in which, for each n, the $n - jd$ is written below n, the d to the right of $n - jd$; and the j along the arrow connecting n to d. The bottom row of Fig. 29.1 now gives rise to the trees in Fig. 29.2 (roots drawn on top). By joining these according to the line above in Fig. 29.1, we

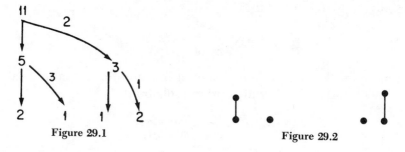

Figure 29.1   Figure 29.2

Figure 29.3   Figure 29.4

obtain Fig. 29.3. Finally, join these again, and obtain Fig. 29.4. Algorithm RANRUT is based on the idea of constructing the tree in Fig. 29.1 in parts, going downward whenever possible, and to the right only when needed. Then, we save what can be constructed and combine pieces as quickly as possible. Figure 29.5 shows, in each column, the part of Fig. 29.1 that has at any moment been determined, and under it a list of relevant graphs from Figs. 29.2–4, which are being constructed at the same time. Each column constitutes one step of progress over the previous one. The thin arrows indicate transitions. The numerical codes 1 and 2 are translated into trees because there is only one tree with that number of vertices. Pairs (j, d) are split up into a pair $(j, 0)$ which serves as a reminder on how to combine two trees once they have been constructed, and d which indicates the size of the tree which is to be worked on next.

The pairs of integers (j, d) are stored in one list and are retrieved as needed, on a "last in, first-out" basis (a so-called "stack"). The partial graphs are stored similarly, and constitute a second stack. The graph is finished when the first stack is empty; the tree T just constructed is the desired output.

An examination of the Figs. 29.1–4 indicates that the stack of pairs of integers never contains more than n elements. Similarly, the stack of trees contains a total of no more than n vertices.

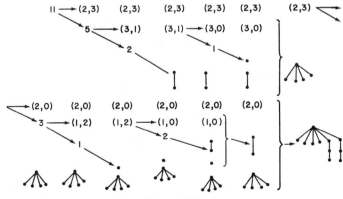

Figure 29.5

Comment. The notion of unlabeled rooted tree means "equivalence class under isomorphisms of trees in which roots correspond to roots." In more practical terms, the present algorithm claims to produce each equivalence class of rooted n-trees with equal probability, but that does not necessarily hold for each of the inevitable labelings of the vertices in a computer output with which a tree of one type can occur. It is in that sense that the trees here are unlabeled even though the computer forces each one to carry a labeling.

We now describe the implementation of these ideas in a computer program. Two stacks are used. The first of these, STACK, will hold the pairs (j, d) or $(j, 0)$, and its elements are therefore ordered pairs of integers.

The second stack, called TREE, is a stack of rooted trees. Initially, TREE is empty. On output, TREE holds the output random rooted tree, in the form: (I,TREE(I)) is the Ith edge of the output tree (I=2,N). At an intermediate stage of the calculation TREE will hold a number of rooted trees, and two kinds of information are stored in TREE: if I is not the root of one of the rooted trees, then (I,TREE(I)) is an edge of such a tree; if I is the root of such a tree then TREE(I) points to the root of the next (smaller I) tree in the stack.

For example, suppose we had, at some stage, the following three rooted trees:

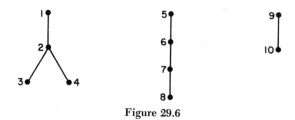

Figure 29.6

Then the portion of the array TREE which would so far be filled would look like this:

I	1	2	3	4	5	6	7	8	9	10	$\cdots$
TREE(I)	0	1	2	2	1	5	6	7	5	9	$\cdots$

One additional pointer, called L in the program, is used to point to the root of the rightmost tree in the list (in the example above, L=9).

In terms of this array TREE, the reader should observe the simplicity of carrying out the two operations which are demanded by the program: adding a new tree to the stack, or performing the operation $T \leftarrow T' + j \otimes T''$, both of which are done in the TREE array itself. The formal algorithm follows.

ALGORITHM RANRUT

[*Note:* T_1, T_2 denote, respectively, the unique rooted unlabeled trees on 1, 2 vertices; Input is NN, output is a random rooted tree on NN vertices.]

(A) $n \leftarrow$ NN; Set STACK and TREE arrays empty.

(B) If $n \le 2$, then write a T_n on the list TREE and go to (C); Else, find a pair (j, d) by (5) and write (j, d) on STACK; Set $n \leftarrow n - jd$ and go to (B).

(C) Read in a pair (j, d) from STACK; If $d = 0$ then join j copies of the last tree in TREE to the root of the next-to-last tree in TREE leaving the resulting tree in TREE to replace both of them; If STACK is empty, exit with the output tree in TREE; else go to step (C). Else, if $d \ne 0$, write $(j, 0)$ on STACK, set $n \leftarrow d$ and go to (B) ∎

In the FORTRAN program, the instructions prior to number 10 simply calculate the numbers $t_1, t_2, \ldots, t_n$ from the recurrence (4). Instructions 10–60 choose a pair of integers (j, d) according to the probabilities (5) and write the pair on STACK.

Step (B) of the formal algorithm begins at instruction 70 where $n \le 2$ and the new tree is placed on TREE and linked to its left neighbor.

Step (C) of the formal algorithm begins at instruction 90. The stack counter IS1 is decremented as the next (j, d) is read in. Then in the DO 104... loop we make J copies of the last tree and write them in TREE, taking care that their roots are all set to the root of the next-to-last tree (this root is called LL, and the IF (MOD ...) instruction sets every Mth array element, i.e., the root of each of the j copies, to LL).

SUBROUTINE SPECIFICATIONS

(1) *Name of subroutine:* RANRUT.

(2) *Calling statement:* CALL RANRUT(NN,T,STACK,TREE).

(3) *Purpose of subroutine:* Generate random unlabeled rooted tree.

(4) *Descriptions of variables in calling statement:*

Name	Type	I/O/W/B	Description
NN	INTEGER	*I*	Number of vertices in desired tree.
T	INTEGER(NN)	*B*	T(I) is the number of rooted, unlabeled trees of I vertices (I=1,2,...) (universal constants).
STACK	INTEGER(2,NN)	*W*	Working storage.
TREE	INTEGER(NN)	*O*	(I,TREE(I)) is the Ith edge of the output tree (I=2,NN), TREE(1)=0.

(5) *Other routines which are called by this one:* FUNCTION RAND(I) (random numbers).

(6) *Number of* FORTRAN *instructions:* 63.

```
      SUBROUTINE RANRUT(NN,T,STACK,TREE)
      IMPLICIT INTEGER(A-Z)
      REAL RAND
      DIMENSION TREE(NN),STACK(2,NN),T(NN)
      DATA NLAST/1/
      L=0
      T(1)=1
1     IF(NN.LE.NLAST) GO TO 10
      SUM=0
      DO 2  D=1,NLAST
      I=NLAST+1
      TD=T(D)*D
      DO 3  J=1,NLAST
      I=I-D
      IF(I.LE.0) GO TO 2
3     SUM=SUM+T(I)*TD
2     CONTINUE
      NLAST=NLAST+1
      T(NLAST)=SUM/(NLAST-1)
      GO TO 1
10    N=NN
      IS1=0
      IS2=0
12    IF(N.LE.2) GO TO 70
20    Z=(N-1)*T(N)*RAND(1)
      D=0
```

```
30    D=D+1
      TD=D*T(D)
      M=N
      J=0
40    J=J+1
      M=M-D
      IF(M.LT.1) GO TO 30
50    Z=Z-T(M)*TD
      IF(Z.GE.0) GO TO 40
60    IS1=IS1+1
      STACK(1,IS1)=J
      STACK(2,IS1)=D
      N=M
      GO TO 12
70    TREE(IS2+1)=L
      L=IS2+1
      IS2=IS2+N
      IF(N.GT.1) TREE(IS2)=IS2-1
80    N=STACK(2,IS1)
      IF(N.EQ.0) GO TO 90
      STACK(2,IS1)=0
      GO TO 12
90    J=STACK(1,IS1)
      IS1=IS1-1
      M=IS2-L+1
      LL=TREE(L)
      LS=L+(J-1)*M-1
      IF(J.EQ.1) GO TO 105
      DO 104  I=L,LS
      TREE(I+M)=TREE(I)+M
      IF(MOD(I-L,M).EQ.0) TREE(I+M)=LL
104   CONTINUE
105   IS2=LS+M
      IF(IS2.EQ.NN) RETURN
      L=LL
      GO TO 80
      END
```

SAMPLE OUTPUT

Subroutine RANRUT was called 450 times with NN=5. There are 9 different rooted unlabeled trees of 5 vertices (Fig. 29.7). The

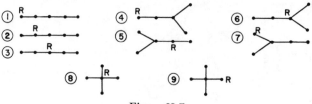

Figure 29.7

frequency with which each of these 9 trees was constructed by RANRUT is shown in the output below.

The value of χ^2 (not shown) is $\chi^2 = 5.6$ with 8 degrees of freedom. In 95% of such experiments the value of χ^2 would lie between 2.03 and 18.17 if the choice of the trees were truly uniform.

```
1   58
2   48
3   44
4   52
5   52
6   50
7   40
8   58
9   48
```

30

Tree of Minimal Length (MINSPT)

Suppose we are given n cities, which are to be interconnected by a communications network by connecting certain pairs of cities, and suppose the network is to be as short as possible. For example, in Fig. 27.5 on page 265 there appears a map of five cities along with the distances between each pair of them. How can we find the shortest network, shown in Fig. 27.6 of the same page, without examining all possibilities?

It happens that this is one of the pleasant mathematical situations in which a method which is as "greedy" as possible at each step turns out also to be optimal. The algorithm which we describe in this chapter will deliver the shortest interconnection in $O(n^2)$ running time.

Observe first that in any given interconnecting network, if a closed path exists then an edge can be removed without disconnecting the network. Hence the connection of minimum length will be a tree which spans the n given points (i.e., visits all of them).

The algorithm itself is recursive. Let T_0 denote the tree which consists of the single vertex $\{n\}$. Generically, suppose that a tree T_{j-1} has been constructed. Then adjoin to T_{j-1} a single edge a_j whose length is minimal in the class of all edges with one end in T_{j-1} and one end not in T_{j-1}. This defines T_j $(j = 1, 2, \ldots, n-1)$.

We claim that T_{n-1} is a spanning tree of minimal length.

Indeed, suppose T^* is any spanning tree of minimal length. Let D^*, D be the lengths of T^*, T_{n-1}, respectively. Suppose $D > D^*$. Let a_i be the first edge in the sequence

$$a_1, a_2, \ldots, a_{n-1}$$

of edges of T_{n-1}, which does not appear in T^*. In T^*, let b denote the edge which joins the component C_1 spanned by edges $a_1, \ldots, a_{i-1}$ to the component of T^* induced by the vertices of T^* which are not in C_1. Suppose b is longer than a_i. Then we could replace b by a_i and obtain a shorter tree, contradicting the minimality of T^*. Suppose b is shorter than a_i. Then we would have chosen b instead of a_i at the ith stage of our construction of T_{n-1}. Hence b and a_i have the same length.

Replace b by a_i in T^*, which leaves its length invariant, and repeat the argument. The process halts after at most $n-1$ steps with T^* having been transformed into T_{n-1} by a sequence of length-preserving edge substitutions. ∎

For the implementation of the algorithm we suppose that input to the subroutine will consist of n and the distance matrix dist(i, j) $(i, j = 1, n)$.

Output will be a single linear array tree(i) $(i = 1, n-1)$ such that $(i, \text{tree}(i))$ is the ith edge of a minimal spanning tree $(i = 1, n-1)$.

At a typical intermediate stage of the algorithm, the array tree(i) will hold two kinds of numbers. A negative entry tree$(i) = -m$ indicates that vertex i is not in the current partially constructed spanning tree, and furthermore the vertex which *is* in the current partial tree which is closest to vertex i is vertex m. A positive entry tree$(i) = m$ indicates that vertex i is in the current tree, and indeed, edge (i, m) is in the tree.

Initially this array is set to $(-n, -n, \ldots, -n, 0)$, corresponding to an initial tree $T_0 = \{n\}$. The full algorithm follows.

ALGORITHM MINSPT

(A) Set tree$(i) \leftarrow -n$ $(i = 1, n-1)$, then do step (B) $n-1$ times and exit.

(B) Let

$$d_{\min} \leftarrow \min\{\text{dist}(i, |\text{tree}(i)|) \mid 1 \leq i \leq n-1; \text{tree}(i) < 0\}$$

and let $i_{\min}$ be a value of i at which the minimum is attained;

[Adjoin new edge] tree(i_{min}) ← −tree(i_{min}); [*Update list of nearest vertices*]

For $i = 1, n - 1$ *do:*

If tree(i) < 0 and dist(i, i_{min}) < dist($i, -$tree(i))

set tree(i) ← −i_{min}

End ■

SUBROUTINE SPECIFICATIONS

(1) *Name of subroutine:* MINSPT.
(2) *Calling statement:* CALL MINSPT(N,NDIM,DIST,TREE).
(3) *Purpose of subroutine:* Find spanning tree of minimal length.
(4) *Descriptions of variables in calling statement:*

Name	Type	I/O/W/B	Description
N	INTEGER	I	Number of vertices.
NDIM	INTEGER	I	Dimension of DIST in calling program.
DIST	REAL(N,N)	I	DIST(I,J) = distance from I to J ($1 \leq I$, $J \leq N$).
TREE	INTEGER(N)	O	(I,TREE(I)) is the Ith edge of the output minimal tree (I=1,N−1).

(5) *Other routines which are called by this one:* None.
(6) *Number of* FORTRAN *instructions:* 24
(7) *Remarks:* If input graph is not complete, put DIST= ∞ on missing edges.

```
      SUBROUTINE MINSPT(N,NDIM,DIST,TREE)
      REAL DIST(NDIM,NDIM)
      INTEGER TREE(N)
      NM1=N-1
20    DO 21  I=1,NM1
21    TREE(I)=-N
      TREE(N)=0
25    DO 51  L=1,NM1
      DMIN=1.E50
40    DO 41  I=1,NM1
      IT=TREE(I)
      IF(IT.GT.0) GO TO 41
```

```
       D=DIST(-IT,I)
       IF(D.GE.DMIN) GO TO 41
       DMIN=D
       IMIN=I
41     CONTINUE
       TREE(IMIN)=-TREE(IMIN)
50     DO 51 I=1,NM1
       IT=TREE(I)
       IF(IT.GT.0) GO TO 51
       IF(DIST(I,IMIN).LT.DIST(I,-IT)) TREE(I)=-IMIN
51     CONTINUE
       RETURN
       END
```

SAMPLE OUTPUT

We present two sample problems. One of them appears in Chapter 27(**E**), page 265, where a network appears, along with its minimal spanning tree.

For a somewhat larger example, in Fig. 30.1 we show a minimal

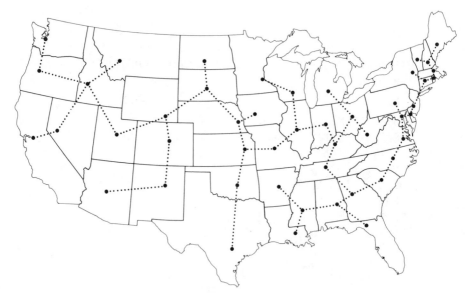

Figure 30.1

spanning tree for the capitols of the 48 contiguous states of the United States. The complete connection which is displayed there would require about 8850 miles of telephone cable. The output was obtained by reading in the latitudes and longitudes of each state capitol, calculating the 48×48 DIST matrix, and then calling MINSPT. The running time was 0.6 seconds with a WATFIV compiler on an IBM 370/168 computer.

Exercises

[The numbers in brackets are those of the relevant chapters.]

1. Find the first ten Taylor series coefficients, about the origin, of $w(z)$, the solution of the equation $we^{-w} = z$. [21]

2. Statistically estimate, by 1000 random trials, the probability that a shuffled deck of 52 cards contains no "straight," i.e., a set of five consecutive cards of consecutive face values (Ace $= 1$, Jack $= 11$, Queen $= 12$, King $= 13$). [8]

3. Tabulate the number of labeled trees on n vertices which have exactly j endpoints ($j = 1, 2, \ldots , n$; $n = 2, 3, 4, 5, 6$). [27(E)]

4. Find the average number of endpoints in a sample of 200 random labeled trees on n vertices ($n = 2, 3, 4, 5, 6$). [28]

5. Deal poker hands to k people. [4]

6. The symmetric group S_n is generated by just *two* elements

$$t : 1 \to 2; 2 \to 1; 3 \to 3; \ldots ; n \to n$$

$$u : 1 \to 2 \to 3 \to \cdots \to n \to 1$$

Sequence the 24 elements of S_4 so that each is obtained from its predecessor by either t or u. S_5 cannot be so sequenced. Prove this by exhaustive computation. [7, 27]

7. For some small values of n, estimate by random trials the average number of edges which must be added one at a time to the totally disconnected graph on n vertices in order to connect it. [18]

8. Given A, an $n \times n$ integer matrix. Output a single integer word whose ith bit position is 1 or 0 according to whether the ith row sum of A is odd or even, respectively.

9. For Hamilton's graph of Fig. 27.4,

    (a) calculate the chromatic polynomial $P(\lambda; G)$
    (b) evaluate $P(\lambda; G)$ $(\lambda = 0(1)8)$
    (c) find the edge-connectivity of G
    (d) what is the chromatic number χ of G?
    (e) list all χ-colorings of G. [19, 20, 22, 27(b)]

10. Modify PERMAN to calculate the permanent of a double-precision complex matrix. Test your program to make sure it works. [23]

11. (a) Which of the "random" routines in this book do you think would provide a sensitive test of the randomness of a random number generator? Discuss.

    (b) Use the program of your choice to compare any three random number generators.

12. Think about how a subroutine can discover how many bits are in a machine word in the machine on which the subroutine is being run.

13. Given blocks of information B(1),...,B(N) (each block could be, e.g., a row of a matrix). Assume a subroutine MOVE(I,J) will move the content of B(I) to B(J). There is one more block B0. Information is moved into and out of B0 by calling MOVE(I,0) or MOVE(0,I). [16]

    Permute the blocks so that B(P(I)) is moved to B(I) (I=1,N), where P is a given permutation.

14. (a) Identify the revolving-door algorithm as the unfolding of a Hamilton circuit on a certain graph $G(n, k)$. Describe $G(n, k)$. [3]

    (b) How many vertices does $G(n, k)$ have? How many edges?
    (c) What is the valence of each vertex of $G(n, k)$?

(d) Draw $G(5, 3)$.

(e) Use HAMCRC [27(D)] to list all of the Hamilton circuits of $G(5, 3)$. Identify the RD algorithm as one of these circuits.

15. (a) Determine if two given graphs G, H are isomorphic. [27]

(b) Determine all isomorphism types of graphs on six vertices.

16. Sample a nonnegative integer n at random from the Poisson distribution

$$\mathrm{Prob}(n) = \frac{a^n}{n!}\, e^{-a} \quad (n = 0, 1, \ldots)$$

where $a > 0$ is given. (See also [K1, Vol. 2].)

17. Determine if the integers from 1 to n, inclusive, can be arranged in three sets so that if a and b are in the same set, then $a + b$ is not in that set. Run this program to find the largest n for which it can be done. [27]

18. Given a Boolean polynomial f of M terms, in N variables. Determine if f is a tautology (always true), and if not, output a set of values of the variables at which f is false. [1]

19. Given two partitions of n: π and π'. Output $+1$ if π is a refinement of π', -1 if π' is a refinement of π, 0 otherwise. [9]

20. Write a program that will find the Möbius function of the set of partitions of n, partially ordered by refinement. [9, 22]

21. Given a permutation π on n letters, construct its inversion table. Try to do it in $O(n \log n)$ steps. [7]

22. (a) Output the list of edges of a random graph on n labeled vertices, k edges, without loops or multiple edges.

(b) Estimate, by random trials, the probability that such a graph is connected, for some small values of k, n. [4, 18]

23. Show that when the chromatic polynomial is expressed in factorial form, $m!$ times the coefficient of $(\lambda)_m$ is the number of colorings in *exactly* m colors, $m = 0, 1, \ldots$. [19, 20]

24. To find random subsets of $\{1, \ldots, n\}$, note that 2^{-i} is the probability that i is the smallest element. A random ξ can determine, without a search, a_1 as the smallest i for which $2^{-1} \leq \xi$ (use logarithms!); adjoin this i to the subset; now work on $\{i + 1, \ldots, n\}$, etc. Work out all details and estimate the labor involved per subset. [2]

25. What is a rhyme scheme, in poetry? How many rhyme schemes can an n-line poem have? Print all possible rhyme schemes of an n-line poem, $n = 2, 3, 4, 5, 6$. (Our thanks to John Riordan for this one.) (See also [Ga2].) [11]

26. (a) Construct the vertex-adjacency matrix of the n-cube Q_n.
(b) List the Hamilton circuits of Q_4. [1]

27. Print the multiplication table of S_n. [7]

28. To choose random k-subsets of $\{1, \ldots, n\}$: for each $i = 1, \ldots, k$, choose l at random in $[1, n - i + 1]$; let a_i be the lth smallest of the integers in $\{1, \ldots, n\} - \{a_1, \ldots, a_{i-1}\}$.

(a) Prove that the method works.
(b) Estimate its operation count.
(c) Modify it by binary search and insertion to obtain the elements in increasing order. [4]

29. Given numbers $k_1, \ldots, k_l$ whose sum is n. Write a subroutine which randomly partitions $\{1, \ldots, n\}$ into subsets $S_1, \ldots, S_l$ with $k_1, \ldots, k_l$ elements, respectively.
Can you do this with $O(nl)$ operations? With $O(n \log l)$ operations? [12]

30. Let $S(n_1, \ldots, n_m)$ be a multiset, which contains n_i copies of $i (i = 1, \ldots, m)$. Design a "NEX" algorithm which produces each of the submultisets of S (all, or those of a fixed cardinality) exactly once.

31. Consider a tree on n vertices, rooted at 1. Determine for each vertex its distance from the root. Can you do this in $O(n)$ time?

32. Let $a_1, \ldots, a_k$ be given positive integers. For each $n \geq 0$, let $P(n)$ denote the number of representations of n in the form

$$n = \mu_1 a_1 + \mu_2 a_2 + \cdots + \mu_k a_k \quad (\mu_i \geq 0, \forall i)$$

(a) Show that

$$\frac{1}{(1 - x^{a_1}) \cdots (1 - x^{a_k})} = \sum_{j=0}^{\infty} P(j) x^j$$

(b) Differentiate logarithmically to show that

$$nP(n) = \sum_{m<n} \tilde{\sigma}(n - m) P(m) \quad (P(0) = 1)$$

Describe $\bar{\sigma}(m)$.

(c) Modify RANPAR to select a representation of n in this form at random (see also [NW2]).   [10]

33. Transpose an $m \times n$ matrix in place using no additional array storage.

34. Devise an algorithm which will generate all compositions of n into k parts, sequenced so that each is obtained from its immediate predecessor by a single jump of one ball from one cell to another.
                                                                 [5]

35. Devise an algorithm which will generate all partitions of an n-set, sequenced so that each is obtained from its predecessor by changing the class of some single element.   [11]

36. Given an $m \times n$ matrix over the integers modulo 2, with $m \leq 4$. Devise an algorithm which counts the nonsingular $m \times m$ submatrices, in less than $O(n^m)$ time. Can you do it in $O(n)$ time?

37. Write a FORTRAN subroutine of ≤ 25 instructions, which will select at random a binary tree on n vertices and output the edge list.

38. (a) Given an array B(1),...,B(N). Describe a method of sorting B into nondecreasing order by constructing the zeta matrix of a certain partial order, then calling TRIANG, then calling RENUMB.

(b) Why is this a very inefficient sorting method? [15, 17, 25]

39. Tabulate the number of displacements and the number of comparisons required to sort each of 100 random input vectors B(1),...,B(N). Print the average and the maximum for N=3,4,...,12. Do this for the Heapsort program of Chapter 15, and for two other sorting methods of your choice.   [8, 15]

40. Let $F(n, j)$ denote the number of permutations of n letters such that the Heapsort program requires exactly j exchanges of pairs in order to sort the permutation into ascending order.

Tabulate $F(n, j)$ for $n = 2, 3, 4, 5, 6, 7$ and all j.   [15, 7]

41. Encode the elements of the field $GF(p^k)$ (p prime, $k \geq 1$) of p^k elements, and construct routines (or tables) for the 4 basic operations.

42. Given long lists $L_1, \ldots, L_k$ of integers. Suppose each L_i is sorted in increasing order. Merge these lists into one sorted list, using no more than $O(\log k)$ comparisons for each integer in the list.

Try to do it in *one* pass. (This is of importance, e.g., if $L_1, \ldots, L_k$ fill all available disk space, and only one tape is available for output.)

43. Given a generalized chessboard; i.e., a finite subset of the squares of a chessboard with infinitely many squares in all directions. Determine if it can be covered by dominoes, and if so, find such a covering. [*Hint:* the black and white squares form a bipartite graph.] [22]

44. Given a partially ordered set with n elements. Devise a "NEX" algorithm which produces all the labelings of the elements with the integers $1, \ldots, n$, so that the ordering of the labels is a refinement of the given partial order. [25]

45. Suppose someone wished to make RANPER "more" random, by replacing step (**B**) by: For $m = 1, n$: $\{p \leftarrow 1 + [\xi n]$; Exchange $a_p, a_m\}$. Prove that this does not work. [8]

46. Write an algorithm which will output the Prüfer word of a labeled tree in linear time. [28]

47. [Refer to Problem 44] Identify the consistent labelings of the elements of a partially ordered set as the set of walks or a certain graph [*Hint:* Generalize the identification of Young Tableaux as such a family of walks in Chapter 14.] [13, 14]

48. Take algorithm NEXT of Chapter 13 in the case of partitions of a set of n elements into k classes, and restate it directly in the language of set-partitions (i.e., making no reference to walks, graphs, codewords, etc.). [13]

49. Devise *encoding* algorithms for the combinatorial families of Chapter 13 (e.g., given a partition of a set in the "usual" format, find the walk to which it corresponds). [13]

50. Identify the recurrence for Fibonacci numbers as a recurrence for the number of objects of order n in a family of walks on the positive integers.

Deduce a unique representation theorem for positive integers as sums of Fibonacci numbers from the fact that every object in the family has a unique rank. [13]

51. Program the first algorithm for random partitions which is mentioned in Chapter 10. [10]

52. Write a program which performs the functions of SPANFO as a depth first search and which uses BACKTR. [18, 27]

Bibliographic Notes

Chapter 1   This algorithm is well known. See Gilbert [G1] for further interesting properties of paths on the cube. For a history of the Gray Code, see Gardner [Ga1].

Chapter 4   Random number generators are in Ralston and Wilf [RW1] and Knuth [K1, Vol. II]. The method of this chapter is announced in Nijenhuis [N1].

Chapter 5   Three other algorithms are mentioned in Lehmer [L1].

Chapter 7   Many methods are known for generating permutations. For Wells's see [W1] or his book [W2]. The adjacent mark method is due to Trotter [Tr1]. Other possibilities are in Lehmer [L1].

Chapter 9   For combinatorial properties of Stirling Numbers see Feller [F1] and Knuth [K1].

Chapter 10   These ideas are from Nijenhuis and Wilf [NW1].

Postscript   See Bender and Goldman [BG1] and Foata and Schützenberger [FS1].

Chapter 12   For an account of Bell numbers, see Gardner [Ga2].

Chapter 13 This material is from Wilf [Wi2], [Wi3].

Chapter 14 This material is from Greene, Nijenhuis, and Wilf [GNW1].

Chapter 15 A definitive discussion of sorting is in Knuth [K1, Vol. III]. The Heapsort is due to Floyd [Fl1] and Williams [Wl1].

Chapter 18 Tarjan's algorithm is in [Ta1].

Chapter 21 Faa diBruno's formula is discussed in Knuth [K1]. The logarithmic differentiation algorithm for $f(z)^n$ is well known [K1, Vol. II, Section 47]. Its use for a general $g(f(z))$ seems to be new.

Chapter 22 The standard reference on network flows is Ford and Fulkerson [FF1]; see also [Be1]. For an in-depth discussion of edge-connectivity see Tutte [Tu1]. Karzanov's algorithm is in [Ka1]; see especially Even's exposition in [Ev2]. The modified method which replaces a layered structure by the more general *KZ*-net and avoids the use of push-down stacks by observing the role of order-ideals is due to Nijenhuis[N2]. See also [BK1].

Chapter 23 Ryser's formula is in [Ry1]; For the rencontres numbers, see Riordan [Ri1]. The factor of 2, as in Eq. (24), can also be derived from Eq. (2) of Wilf [Wi1]. The applicability of the Gray code has also been observed by Knuth [K1, Vol. 2, p. 440] who also gives an algorithm as fast as ours, which however requires nearly 2^n storage registers.

A short and elementary proof of Bregman's theorem (conjecture of Ryser and Minc[M1], [Ry2]) is in [Sc1].

Chapter 28 The linear-time decoding algorithm is due to Paul Klingsberg (doctoral dissertation, University of Pennsylvania, 1977).

References

Beckenbach, E.
[Be1]   Network flow problems, in "Applied Combinatorial Mathematics" (E. Beckenbach, ed.). Wiley, New York, 1964.
Bender, E. A., and Goldman, J. R.
[BG1]   Enumerative uses of generating functions, *Indiana Univ. Math. J.* **20** (*1971*), 753–765.
Bender, E. A., and Knuth, D. E.
[BK1]   Enumeration of plane partitions, *J. Combinatorial Theory Ser. A* **13** (1972) 40–54.
Conte, S. I., and de Boor, C.
[CB1]   "Elementary Numerical Analysis" 2nd ed. McGraw-Hill, New York, 1972.
Even, S.
[Ev1]   "Combinatorial Algorithms." Macmillan, New York, 1973.
[Ev2]   The max flow algorithm of Dinic and Karzanov. Lecture Notes, Massachusetts Inst. of Tech., Cambridge (1976).
Feller, W.
[F1]   "An Introduction to Probability Theory and Its Applications." Wiley, New York, 1951.
Floyd, R. W.
[Fl1]   *Comm. ACM.* **7** (1964), 701.
Foata, D., and Schützenberger, M.
[FS1]   "Théorie géométrique des polynomes euleriens" (Lecture Notes in Math., No. 138). Springer-Verlag, Berlin and New York, 1970.
Ford, L. R., and Fulkerson, D. R.
[FF1]   "Flows in Networks." Princeton Univ. Press, Princeton, New Jersey, 1962.

Gardner, M.

[Ga1]   *Sci. Amer.* (August 1972), pp. 105–109.

[Ga2]   *Sci. Amer.* (May 1978), pp. 24–30.

Gilbert, E. N.

[G1]   Gray codes and paths on the *n*-cube, *Bell System Tech. J.* 37 (1958), 815–826.

Gilbert, E. N., and Pollak, H. O.

[GP1]   Steiner minimal trees, *SIAM J. Appl. Math.* 16 (1968), 1–29.

Goldman, J. R.

    See [BG1].

Greene, C., Nijenhuis, A., and Wilf, H. S.

[GNW1]   A probabilistic proof of the hook formula; *Advances in Math.* to appear.

Hutchinson, J. P.

[H1]   Eulerian graphs and polynomial identities for sets of matrices, *Proc. Nat. Acad. Sci. U.S.A.*, 1974.

Karzanov, A. V.

[Ka1]   Determining the maximal flow in a network by the method of preflows, *Soviet Math. Dokl.* 15 (1974) 434–437.

Knuth, D.

[K1]   "The Art of Computer Programming" (3 vols.). Addison-Wesley, Reading, Massachusetts, 1968, 1969, 1973.

Kostant, B.

[Ko1]   A theorem of Frobenius, a theorem of Amitsur-Levitzki and cohomology theory, *J. Math. Mech.* 7 (1958) 237–264.

Kruskal, J. B., Jr.

[Kr1]   On the shortest spanning subtree of a graph and the travelling salesman problem, *Proc. Amer. Math. Soc.* 7 (1956), 48–50.

Lehmer, D. H.

[L1]   The machine tools of combinatories, *in* "Applied Combinatorial Mathematics" (E. Beckenbach, ed.). Wiley, New York, 1964.

Liu, C. L.

[Li1]   "Introduction to Combinatorial Mathematics." McGraw-Hill, New York, 1968.

Minc, H.

[M1]   Upper bounds for permanents of (0, 1) matrices, *Bull. Amer. Math. Soc.* 69 (1963), 789–791.

Moon, J. W.

[Mo1]   Various proofs of Cayley's formula for counting trees, *in* "A Seminar on Graph Theory" (L. Beineke and F. Harary, eds.). Holt, New York, 1967.

[Mo2]   Counting labelled trees, *Canad. Math. Monographs*, No. 1, 1970.

Nijenhuis, A.

[N1]   Random subsets of fixed size, with optimal time and storage, *Notices Amer. Math. Soc.* (to appear).

[N2]   Network flow with linear storage requirements, *Notices Amer. Math. Soc.* (to appear).

    See also [GNW1].

Nijenhuis, A., and Wilf, H. S.

[NW1]   A method and two algorithms in the theory of partitions, *J. Combinatorial Theory*, to appear.

[NW2]   Representations of integers by linear forms in nonnegative integers, *J. Number Theory*, 4 (1970), 98–106.

Pollak, H. O.

    See [GP1].

Prim, R. C.
[P1] Shortest connection networks and some generalizations, *Bell System Tech. J.* **36** (1957) 1389–1401.

Ralston, A., and Wilf, H. S.
[RW1] "Mathematical Methods for Digital Computers" (2 vols.). Wiley, New York, 1960, 1966.

Read, R. C.
[R1] An introduction to chromatic polynomials, *J. Combinatorial Theory* **4** (1968), 52–71.

Riordan, John
[Ri1] "An Introduction to Combinatorial Analysis." Wiley, New York, 1958.

Rota, G. C.
[Ro1] On the foundations of combinatorial theory, I. The Möbius function. Z. *Wahrscheinlichkeitstheorie und Verw. Gebiete,* **2** (1964), 340–368.

Ryser, H.
[Ry1] "Combinatorial Mathematics" (Carus Math. Monographs, No. 14). Wiley, New York, 1963.
[Ry2] Matrices of zeros and ones, *Bull. Amer. Math. Soc.* **66** (1960), 442–464.

Schrijver, A.
[Sc1] A short proof of Minc's conjecture, Mathematisch Centrum, Amsterdam, 1977.

Swan, R. G.
[S1] An application of graph theory to algebra, *Proc. Amer. Math. Soc.* **14** (1963), 367–373; Correction, **21** (1969), 379–380.

Tarjan, R. E.
[Ta1] Depth-first search and linear graph algorithms, *SIAM J. Computing I,* 2(1972), 146–160.

Trotter, H.
[Tr1] "PERM", Algorithm 115, *Comm. ACM.* **5** (1962), 434–435.

Tutte, W. T.
[T1] "Connectivity in Graphs." Univ. of Toronto Press, Toronto, 1966.

Walker, R. J.
[Wa1] An enumerative technique for a class of combinatorial problems, in "Combinatorial Analysis," *Proc. Symp. Appl. Math.* **10** (1960), 91–94 (R. Bellman and M. Hall, eds.). Amer. Math. Soc., Providence, Rhode Island.

Wells, M. B.
[W1] Generation of permutations by transposition, *Math. Comp.* **15** (1961), 192–195.
[W2] "Elements of Combinatorial Computing." Pergamon, New York, 1971.

Whitney, H.
[Wh1] The coloring of graphs, *Ann. of Math.* **33** (1932) 688–718.

Wilf, H. S.
[Wi1] A mechanical counting method and combinatorial applications, *J. Combinatorial Theory* **4** (1968), 246–258.
[Wi2] A unified setting for sequencing, ranking and random selection of combinatorial objects, *Advances in Math.* **24** (1977) 281–291.
[Wi3] A unified theory of selection algorithms, II, *Ann. Discr. Math.* **2** (1978) 135–148.
       See also [NW1], [NW2], [RW1], [GNW1].

Williams, J. W. J.
[Wl1] *Comm. ACM.* **7** (1964), 347–348.

Index

Computer Science and Applied Mathematics
A SERIES OF MONOGRAPHS AND TEXTBOOKS

Editor
Werner Rheinboldt
University of Pittsburgh

HANS P. KÜNZI, H. G. TZSCHACH, AND C. A. ZEHNDER. Numerical Methods of Mathematical Optimization: With ALGOL and FORTRAN Programs, Corrected and Augmented Edition

AZRIEL ROSENFELD. Picture Processing by Computer

JAMES ORTEGA AND WERNER RHEINBOLDT. Iterative Solution of Nonlinear Equations in Several Variables

AZARIA PAZ. Introduction to Probabilistic Automata

DAVID YOUNG. Iterative Solution of Large Linear Systems

ANN YASUHARA. Recursive Function Theory and Logic

JAMES M. ORTEGA. Numerical Analysis: A Second Course

G. W. STEWART. Introduction to Matrix Computations

CHIN-LIANG CHANG AND RICHARD CHAR-TUNG LEE. Symbolic Logic and Mechanical Theorem Proving

C. C. GOTLIEB AND A. BORODIN. Social Issues in Computing

ERWIN ENGELER. Introduction to the Theory of Computation

F. W. J. OLVER. Asymptotics and Special Functions

DIONYSIOS C. TSICHRITZIS AND PHILIP A. BERNSTEIN. Operating Systems

A. T. BERZTISS. Data Structures: Theory and Practice, Second Edition

N. CHRISTOPHIDES. Graph Theory: An Algorithmic Approach

SAKTI P. GHOSH. Data Base Organization for Data Management

DIONYSIOS C. TSICHRITZIS AND FREDERICK H. LOCHOVSKY. Data Base Management Systems

JAMES L. PETERSON. Computer Organization and Assembly Language Programming

WILLIAM F. AMES. Numerical Methods for Partial Differential Equations, Second Edition

ARNOLD O. ALLEN. Probability, Statistics, and Queueing Theory: With Computer Science Applications

ELLIOTT I. ORGANICK, ALEXANDRA I. FORSYTHE, AND ROBERT P. PLUMMER. Programming Language Structures

ALBERT NIJENHUIS AND HERBERT S. WILF. Combinatorial Algorithms, Second Edition

AZRIEL ROSENFELD. Picture Languages, Formal Models for Picture Recognition

ISAAC FRIED. Numerical Solution of Differential Equations

ABRAHAM BERMAN AND ROBERT J. PLEMMONS. Nonnegative Matrices in the Mathematical Sciences

BERNARD KOLMAN AND ROBERT E. BECK. Elementary Linear Programming with Applications

CLIVE L. DYM AND ELIZABETH S. IVEY. Principles of Mathematical Modeling

This is a volume in
COMPUTER SCIENCE AND APPLIED MATHEMATICS
A Series of Monographs and Textbooks

Editor: WERNER RHEINBOLDT

A complete list of titles in this series appears at the end of this volume.